Qualitative Methodologies in Organization Studies

Qualitative Methodologies in Organization Studies

Volume I: Theories and New Approaches

Foreword by Martyna Śliwa

palgrave
macmillan

Editors
Malgorzata Ciesielska
Teesside University Business School
Teesside University
Middlesbrough, UK

Dariusz Jemielniak
Akademia Leona Koźmińskiego
Warszawa, Poland

ISBN 978-3-319-65216-0   ISBN 978-3-319-65217-7   (eBook)
https://doi.org/10.1007/978-3-319-65217-7

Library of Congress Control Number: 2017960769

Printed on acid-free paper

This Palgrave Macmillan imprint is published by Springer Nature
The registered company is Springer International Publishing AG
The registered company address is: Gewerbestrasse 11, 6330 Cham, Switzerland

With love, to my son,
Max Ciesielski-Lattimer
With love, to my daughter, Alicja Jemielniak-Banasik, and wife, Natalia
Banasik-Jemielniak

Preface

The book is designed with organization studies researchers, including Ph.D.s and students of Doctorate in Business Management who would like to understand the current state of art of qualitative research in organization studies. The book is structured to discuss not only the key methods but also broader research design considerations and cutting-edge approaches. All chapters are based on robust and holistic literature reviews and are prepared by active researchers specializing in the theories and approaches they are discussing, which also enables for more practical considerations.

In this first volume, we offer a range of chapters covering key approaches and theories related to qualitative research. Dariusz Jemielniak and Malgorzata Ciesielska introduce qualitative research in organization studies in Chap. 1. In Chap. 2, Bartosz Sławecki discusses the relationship between sociological paradigms, assumptions about the nature of social reality (ontologies) and the nature of scientific cognition (epistemologies), as well as practical ways of conducting social research (methodologies). In Chap. 3, Przemysław Hensel and Beata Glinka explain the logic of the grounded theory research strategy. They present the conceptual roots of the grounded theory approach and provide guidance on the grounded theory research process. In Chap. 4, Slawomir Magala introduces the concept of visual anthropology and how it can be used to understand the social world around us. In Chap. 5, Davydd Greenwood discusses an action research strategy that combines the expertise and facilitation of a professional social researcher with the knowledge, energy and commitments of local stakeholders in an organization. In Chap. 6, Tony Watson discusses organization ethnography as a research method, a descriptive study of people, a theory-informed and theory-informing analysis of (intensive) fieldwork, and a humanities-style written account and analysis of events in which the researcher-writer has participated. In Chap. 7, David Boje and Nazanin Tourani introduce us to materiality storytelling research and how to utilize it to enrich our understanding of various disciplines.

Yiannis Gabriel discusses the importance of interpretation, reflexivity, and imagination in qualitative research in Chap. 8. This is followed by Angela Mazzetti's discussion on emotions and the exploration of some of the emotional challenges of developing rapport with research participants in Chap. 9. In Chap. 10, Daniela Rudloff raises questions around the accessibility of research and discusses how to lower barriers to participation. Finally, Sylwia Ciuk and Dominika Latusek cover a number of ethical questions and ethical dilemmas that can arise at different stages of the research process (Chap. 11).

Middlesbrough, UK   Malgorzata Ciesielska

Warsaw, Poland   Dariusz Jemielniak

Contents

Notes on Contributors

David Boje is a Regents Professor and Distinguished Achievement Professor in the Management Department, New Mexico State University. He is an international scholar in the areas of storytelling and antenarratives in organizations. He holds an honorary doctorate from Aalborg University and is considered godfather of their Material Storytelling Lab founded by Anete Strand. He is the founder of *Tamara—Journal of Critical Organization Inquiry*. He has written 21 books and 141 journal articles, many in top-tier journals, such as *Management Science, Administrative Science Quarterly, Organization Studies, Human Relations* and *Academy of Management Journal*. It is said that he is the most cited scholar in the College of Business at New Mexico State University.

Malgorzata Ciesielska is Senior Lecturer in Organisation Studies and HRM at Teesside University. Her research interests focus on the cross-section of organizational behavior and innovation management, on particular areas of hybrid organizations, and on diversity and cross-professional collaboration in the tech industry. She specializes in qualitative research and organizational ethnography and is an associate editor of *Tamara—Journal for Critical Organization Inquiry* and *Przegląd Europejski* (European Review).

Sylwia Ciuk is Senior Lecturer in Organisation Studies at Oxford Brookes University. She has carried out qualitative studies in Poland and the UK. Her research interests center around discursive perspectives on organizational change. Recently, she has been exploring the role of language and, in particular, interlingual translation in the processes of change. Ciuk is a member of the European

Group for Organizational Studies and the Standing Conference on Organizational Symbolism. Her work has been published both in English and in Polish.

Yiannis Gabriel is Professor of Organizational Theory at Bath University. Gabriel is known for his work in leadership, management learning, organizational storytelling and narratives, psychoanalytic studies of work, and the culture and politics of contemporary consumption. He has used stories as a way of studying numerous social and organizational phenomena, including leader–follower relations, group dynamics and fantasies, the management of change, innovation and knowledge transfer. He is the author of ten books and numerous articles and maintains an active blog in which he discusses music, storytelling, books, cooking, pedagogy and research outside the constraints of academic publishing (http://www.yiannisgabriel.com/). He is the senior editor of *Organization Studies*. His enduring fascination as a researcher and educator lies in what he describes as the unpredictability and complexity of organizational life.

Beata Glinka, PhD is a professor at the University of Warsaw (Poland) and Head of the Department of Organizational Innovations and Entrepreneurship at the Faculty of Management. Her research interests focus on the cultural context of management, entrepreneurship and the culture of public administration (and its influence on entrepreneurship). She has authored and co-authored over 50 articles, book chapters, and books. In her research, she applies qualitative methods. For research, she was awarded several times, for example, by Fulbright Foundation (2011, Senior Advanced researcher Award) and Polish Science Foundation (2003). Glinka is also a co-editor of the scientific journal *Problemy Zarządzania* (Management Issues).

Davydd Greenwood is Goldwin Smith Professor of Anthropology Emeritus (Cornell University, USA). He is a corresponding member of the Spanish Royal Academy of Moral and Political. His work centers on action research, political economy, ethnic conflict, community and regional development, and neo-liberal reforms of higher education. His ethnographic work has focused on the Spanish Basque Country, Spain's La Mancha region and higher education institutions. Greenwood is the author/co-author of nine books and scores of articles. Since the early 1980s, he has focused on the relationships between action research and higher education reform, writing extensively on this subject. He currently participates in an international network to create democratically organized public universities. He is the co-author with Morten Levin of the recent book, *Creating a New Public University and Reviving Democracy: Action Research in Higher Education* (2016).

Przemyslaw Hensel is a professor at the Faculty of Management, University of Warsaw (Poland). His research interests focus on organizational isomorphism in public administration, transfer of managerial templates between different cultures and regions, and methods for organizational diagnosis and assessment. In 2012, he was awarded the Fulbright Advanced Research Award (Senior); earlier, he received awards from The Rector of the University of Warsaw. He is a co-editor of the scholarly journal *Problemy Zarządzania* (Management Issues).

Dariusz Jemielniak is a full Professor of Management, the Head of the Center for Research on Organizations and Workplaces (CROW) and a founder of the New Research on Digital Societies (NeRDS) group at Kozminski University. His interests revolve around critical management studies, open collaboration projects (such as Wikipedia or F/LOSS), narrativity, storytelling, knowledge-intensive organizations, virtual communities, and organizational archetypes, all studied by interpretive and qualitative methods.

Dominika Latusek is Professor of Management at Kozminski University, Poland. Former Fulbright scholar at Stanford University, she has conducted qualitative field studies in companies in Poland, Germany, Sweden, and Silicon Valley in the USA. Her fields of interest include high-growth venture creation, interorganizational relations and trust within and between organizations.

Slawomir Magala is Professor of Cross-Cultural Management at the Rotterdam School of Management, Erasmus University, and at the Faculty of Management and Social Communication of the Jagiellonian University. He has authored a range of books including: *Cross-Cultural Competence* (2005) (Polish translation, 2011), *The Management of Meaning in Organizations* (2009) and *Class Struggle in Classless Poland* (1982) (Polish translation, 2012). He teaches cross-cultural management and sustainable cultural entrepreneurship.

Angela Stephanie Mazzetti has a passion for qualitative research methods gained from the different perspectives of a researcher, a research supervisor, and a lecturer in qualitative techniques. Mazzetti has written a number of papers focusing specifically on the emotional nature of qualitative research, highlighting some of the challenges for qualitative researchers and on how being in tune with our emotions can contribute to and enhance our research. Mazzetti's research specialism is emotion and stress, and she is keen to explore how qualitative approaches can enhance our knowledge and understanding of how we experience and cope with stress. Mazzetti is currently researching coping resilience and resource depletion, adopting a sensory ethnographic approach.

Daniela Rudloff is a lecturer at the School of Business at the University of Leicester. Her research interests focus on equality and diversity; research methods; and the intersection of the two. She has taught research methods to students at every level, from Certificate to Ph.D. and Doctorate, and has started integrating accessibility into the research methods curriculum. Rudloff is publishing a book on quantitative research methods with Palgrave Macmillan and she is a member of the editorial board of *Sociological Research Online*.

Bartosz Sławecki is an assistant professor in the Department of Education and Personnel Development at Poznań University of Economics and Business in Poland. He holds a Ph.D. in Economics and an M.A. in Sociology and Management. His primary interests include theory and the practice of qualitative research, in particular, the biographical and narrative approaches in organization and labor studies. Currently, his work is focused on identity work and the professional identity development of social entrepreneurs in Poland.

Nazanin Tourani is an assistant professor at Pennsylvania State University at Fayette. Her research interests cover business strategy, human resource management, and organizational behavior. She is specializing in the storytelling approach in organization studies.

Tony Watson is Emeritus Professor of Sociology, Work and Organisation at Nottingham University Business School. His major organizational ethnographic study, *In Search of Management* (originally 1994), is still in print and has sold 11,000 or so copies. His key 'position papers' on organizational ethnography (in the *Journal of Management Studies* and the *Journal of Organizational Ethnography*) are being cited widely. A key organizationally ethnographically study was published in *Human Relations* in 2011.

List of Tables

1

Qualitative Research in Organization Studies

Dariusz Jemielniak and Malgorzata Ciesielska

Qualitative organization studies have never had it easy. Even though they date back to the beginnings of social studies as a whole, and Bronisław Malinowski and Elton Mayo are considered to be one of the founders of anthropology and management and organization studies, the academic status of qualitative inquiry in management has been unequal for quite a while. In the 1960s, when academia became obsessed with cybernetics and the system approach, typically paired with quantitative studies, the perception of qualitative studies was that they are clearly inferior.

Dariusz Jemielniak's work on the publication was possible thanks to a research grant from the Polish National Science Center (no. UMO-2012/05/E/HS4/01498).

D. Jemielniak (✉)
Kozminski University, Warsaw, Poland

M. Ciesielska
Teesside University Business School, Middlesbrough, UK

© The Author(s) 2018
M. Ciesielska, D. Jemielniak (eds.), *Qualitative Methodologies in Organization Studies*,
https://doi.org/10.1007/978-3-319-65217-7_1

1

Fortunately, it soon turned out that there are questions that quantitative studies, in spite of their undisputable usefulness, cannot address well. As the sensitivity to new cultures has grown, so has the interest in studying different cultures, including the organizational ones, and qualitative studies have naturally returned to grace. Currently, we often use qualitative approaches not only in regular organization and occupational studies (Ciesielska 2008; Ciesielska and Petersen 2013; Jemielniak 2002, 2005; Konecki 1990), but also in the studies of knowledge-intensive environments (Bowden and Ciesielska 2016; Ciesielska and Petersen 2013; Latusek-Jurczak and Prystupa 2014) as well as of virtual tribes and the digital society (Jemielniak 2013, 2015; Przegalinska 2015). Qualitative studies are even sometimes positioned as a meso-level model for doing social science in general (Gaggiotti et al. 2016).

What makes qualitative studies "qualitative" though? For some people the category of qualitative studies encompasses everything that cannot be quantified. As such, they would have to include also simple literature reviews or theoretical essays (Jemielniak and Aibar 2016; Jemielniak and Greenwood 2015). They would also, by definition, have to cover all kinds of poorly designed research reports that do not even make a serious attempt at presenting their methodology. On the other side of the spectrum there are qualitative researchers for whom only a deep involvement with the field and saturation of the material can satisfy the rigorous requirements of a qualitative inquiry. As a result, labeling certain tools and approaches as "qualitative" may be discipline-dependent. For instance, some econometrists may consider open questions in a questionnaire as qualitative research, while for anthropologists they could be intrinsically quantitative, and even a loosely structured interview (Whyte and Whyte 1984) can be perceived as "not qualitative enough" to allow a solid interpretation.

Things are further complicated by the fact that research methods are only secondary to some more fundamental methodological and paradigmatic choices. Methods are tools that can be used for different purposes: just like with a hammer one can hit some nails, bring down walls, or paint the walls—although, admittedly, the latter can be done much easier and with better results with a brush. Especially in engaged scholarship and action research (Greenwood and Levin 1998; Strumińska-Kutra 2016), it is relatively common to rely heavily on triangulation of approaches (Konecki 2008b), and remain quite agnostic to deep ideological aversions to particular toolsets. Such aversions are still very much alive in organization studies among some researchers, and the pluralism of approaches is not always valued (Pfeffer 1995; Van Maanen 1995). Interestingly, the qualitative scholars have returned from their exile back to the game with new methodological reflexivity (Jemielniak and Kostera 2010), and seem to be more sensitive to the variety of possible views and research angles.

Qualitative studies are deeply imbued in serendipity (Konecki 2008a) and require remaining open to interpretation. This is why they are often performative, and not ostensive (Czarniawska 2017; Latour 1986): they draw from models and theories generated in and through research, and are not preconceptualized before the fieldwork begins. Coming to the field with preconceptions and a belief that we can understand the studied world better than the people who live it are commonly linked with the drive to consult and "be practical". Sometimes this approach stems from an assumption that a scholar should always try to give unsolicited advice to the studied. While we are generally sympathetic to scholars who want to be useful and wholeheartedly commend engaged scholarship, we also recognize that practitioners quite often know their lore much better than the scholars. Throwing ideas from an ivory tower rarely works (Etzkowitz et al. 2000). Claiming that a discipline can be useful for practitioners is a typical and old legitimization strategy, used as early as in the ancient Greece, where philosophy made the promises of solving military problems. Sociology also made such claims early on, when Durkheim proclaimed taking the issue of poverty and crime head on (Czarniawska-Joerges 1999). Instead, we should remain "reflexive practitioners" (Schön 1983) and observe the social actors' actual practices (Barley and Kunda 2001; Czarniawska 2001), without assuming a privileged know-it-all role. Qualitative methods make it more difficult to assume the role of a consultant also because of their focus on the studied and following their logic. However, the qualitative approaches can become powerful tools for an experienced consultant.

References

Barley, S. R., & Kunda, G. (2001). Bringing Work Back In. *Organization Science, 12*(1), 76–95.

Bowden, A., & Ciesielska, M. (2016). Accretion, Angst and Antidote: The Transition from Knowledge Worker to Manager in the UK Heritage Sector in an Era of Austerity. In *The Laws of the Knowledge Workplace: Changing Roles and the Meaning of Work in Knowledge-Intensive Environments*. London: Routledge.

Ciesielska, M. (2008). From Rags to Riches – A Fairy Tale Or A Living Ethos? Stories of Polish Entrepreneurship During and After the Transformation of 1989. In M. Kostera (Ed.), *Organizational Olympians: Heroes and Heroines of Organizational Myths* (pp. 59–70). Basingstoke: Palgrave.

Ciesielska, M., & Petersen, G. (2013). Boundary Object As a Trust Buffer. The Study Of an Open Source Code Repository. *Tamara Journal of Critical Organisation Inquiry, 11*(3), 5.

Czarniawska, B. (2001). Having Hope in Paralogy. *Human Relations, 54*(1), 13–21.

Czarniawska, B. (2017). Organization Studies As Symmetrical Ethnology. *Journal of Organizational Ethnography, 6*(1), 2–10.

Czarniawska-Joerges, B. (1999). *Writing Management: Organization Theory as a Literary Genre*. Oxford/New York: Oxford University Press.

Etzkowitz, H., Webster, A., Gebhardt, C., & Terra, B. R. C. (2000). The Future of the University and the University of the Future: Evolution of Ivory Tower to Entrepreneurial Paradigm. *Research Policy, 29*(2), 313–330.

Gaggiotti, H., Kostera, M., & Krzyworzeka, P. (2016). More than a Method? Organisational Ethnography as a Way of Imagining the Social. *Culture and Organization, 23*(5), 325–340.

Greenwood, D. J., & Levin, M. (1998). *Introduction to Action Research: Social Research for Social Change*. Thousand Oaks: Sage.

Jemielniak, D. (2002). Kultura – odkrywana czy konstruowana? *Master of Business Administration, 2*(55), 28–30.

Jemielniak, D. (2005). Kultura – zawody i profesje. *Prace i Materiały Instytutu Studiów Miedzynarodowych SGH, 32*, 7–22.

Jemielniak, D. (2013). Netnografia, czyli etnografia wirtualna – nowa forma badań etnograficznych. *Prakseologia, 154*, 97–116.

Jemielniak, D. (2015). Naturally Emerging Regulation and the Danger of Delegitimizing Conventional Leadership: Drawing on the Example of

Wikipedia. In H. Bradbury (Ed.), *The SAGE Handbook of Action Research*. London/New Delhi/Thousand Oaks: Sage.

Jemielniak, D., & Aibar, E. (2016). Bridging the Gap Between Wikipedia and Academia. *Journal of the Association for Information Science and Technology*, *67*(7), 1773–1776.

Jemielniak, D., & Greenwood, D. J. (2015). Wake Up or Perish: Neo-Liberalism, the Social Sciences, and Salvaging the Public University. *Cultural Studies – Critical Methodologies, 15*(1), 72–82.

Jemielniak, D., & Kostera, M. (2010). Narratives of Irony and Failure in Ethnographic Work. *Canadian Journal of Administrative Sciences, 27*(4), 335–347.

Konecki, K. (1990). Dependency and Worker Flirting. In B. A. Turner (Ed.), *Organizational Symbolism* (pp. 55–66). Berlin/New York: Gruyter.

Konecki, K. (2008a). Grounded Theory and Serendipity. Natural History of a Research. *Qualitative Sociology Review, 4*(1), 171–188.

Konecki, K. (2008b). Triangulation and Dealing with the Realness of Qualitative Research. *Qualitative Sociology Review, 4*(3), 7–28.

Latour, B. (1986). The Powers of Association. In J. Law (Ed.), *Power, Action and Belief – A New Sociology of Knowledge?* London/Boston/Henley: Routledge & Kegan Paul.

Latusek-Jurczak, D., & Prystupa, K. (2014). Collaboration and Trust-Building in Open Innovation Community. *Journal of Economics & Management, 17*, 47.

Pfeffer, J. (1995). Mortality, Reproducibility, and the Persistence of Styles of Theory. *Organization Science, 6*(6), 681–686.

Przegalinska, A. (2015). Embodiment, Engagement and The Strength Virtual Communities: Avatars of Second Life in Decay. *Tamara, 13*, 48–62.

Schön, D. (1983). *The Reflexive Practitioner. How Professionals Think in Action.* New York: Basic Books.

Strumińska-Kutra, M. (2016). Engaged Scholarship: Steering Between the Risks of Paternalism, Opportunism, and Paralysis. *Organization, 23*(6), 864–883.

Van Maanen, J. (1995). Fear and Loathing in Organization Studies. *Organization Science, 6*(6), 687–692.

Whyte, W. F., & Whyte, K. K. (1984). *Learning from the Field: A Guide from Experience.* Beverly Hills: Sage.

2

Paradigms in Qualitative Research

Bartosz Sławecki

2.1 Introduction

The aim of the chapter is to raise novice researchers' awareness of the significance of philosophical assumptions for their practical activity. The text presents the basic terms connected with the methodology of social sciences. The entire discussion is centered on the issue of paradigms. Various approaches within the framework of basic philosophical assumptions are discussed—concerning the nature of social reality (ontologies), the nature of scientific cognition (epistemologies), and practical ways of conducting social research (methodologies). An important element of the text is the presentation of two classifications of paradigms in social sciences with particular consideration given to qualitative research.

This work was supported by the Polish National Science Center grant 2013/11/D/HS4/03878

B. Sławecki (✉)
Poznań University of Economics and Business, Poznań, Poland

© The Author(s) 2018
M. Ciesielska, D. Jemielniak (eds.), *Qualitative Methodologies in Organization Studies*,
https://doi.org/10.1007/978-3-319-65217-7_2

2.2 Paradigms in Social Sciences

The term "paradigm" is usually associated with the figure of Thomas Kuhn, an American historian and philosopher of science, who died in 1996, and his most famous book, entitled *The Structure of Scientific Revolutions*. Kuhn's work concerns a certain synthetic view on the history of science. Let us add that it is dedicated to natural history, which has significant implications for the shape of the presented concept of the development of science. According to Kuhn, the development of science as a separate field of human activity is discontinuous in nature and, broadly speaking, consists in the alternate occurrence of periods of so-called normal science and scientific revolutions. The term **normal science** refers to "research firmly based upon one or more past scientific achievements, achievements that some particular scientific community acknowledges for a time as supplying the foundation for its further practice" (Kuhn 1970, p. 10). Therefore, the period in which normal science prevails consists in scientists acting in accordance with generally accepted rules without the need to ponder over them or give consideration to the correctness of the adopted assumptions. Giddens (1993) believes that in Kuhn's perspective, the success of the development of science actually depends on suspending "critical reasoning", on accepting the set of basic philosophical assumptions concerning the manner of getting to know the world as obvious.

Scientific revolution means breaking with the only generally shared manner of practicing science; it means moving from one universally accepted manner of studying reality to another, resulting in the emergence of a new tradition of institutional science. As Kuhn claims (1970, p. 111), "during revolutions scientists see new and different things when looking with familiar instruments in places they have looked before. It is rather as if the professional community had been suddenly transported to another planet where familiar objects are seen in a different light and are joined by unfamiliar ones as well. Of course, nothing of quite that sort does occur: there is no geographical transplantation".

According to Kuhn, in a period of normal science, scientists function within a framework of a certain **paradigm**, that is, within a framework of established and socially accepted views, ways of arriving at solutions to scientific problems (scientific methods), cultivated and validated principles and rules of conducting scientific research. As stated by Kuhn, the term is supposed to "suggest that some accepted examples of actual scientific practice – examples which include law, theory, application, and instrumentation together – provide models from which spring particular coherent traditions of scientific research" (ibid., p. 10). A scientific revolution means a departure from a paradigm and its replacement with a new one. A revolution leads to a change of standards and values, a change in the manner of perceiving the world, resulting in the discerning of new elements of the world, not previously researched.

The **social and psychological origin** of a paradigm is an extremely important issue. A paradigm—as a prevalent manner of practicing science—emerges from among various visions, approaches, traditions, ideas, or concepts, as at a certain time, most scientists consider certain assumptions or scientific methods as correct and exemplary. As a particular paradigm becomes recognized as applicable and valid, it is accompanied by a social process of its validation, inheritance, and scientific socialization of the next generation of scientists, which occurs in the course of research work acquiring education. The period of validity of a specific paradigm is therefore relatively stable and isolated from external influences. As Kuhn claims, any external interferences come to the fore only during phases of revolutionary changes.

And so, in a period of the prevalence of a particular paradigm, there occurs a more or less conscious internalization of certain rules of conduct in science, of practicing it in a certain way, in the image and likeness of shared manners of perceiving and analyzing reality. A researcher can even be completely unaware of their entanglement in a generally accepted system of philosophical and methodological principles. Sometimes, paradigms remain deeply hidden; they are silently assumed and treated as obvious by many. Moreover, a researcher may be unaware of the existence of different ways of perceiving and studying the world, which often causes their indifference, incomprehension, or even lack of acceptance for other research orientations—and consequently, hinders and sometimes also prevents cooperation of scientists within a single scientific discipline (cf. Kostera 2007).

A paradigm contains a certain immanent contradiction concerning its **utility**. On the one hand, functioning within a framework of a specific paradigm in a way frees scientists from the need to ponder over the philosophical assumptions concerning the world and the bases of scientific activity which justify and legitimize their actions. This characteristic determines the power of a paradigm—it is useful, because it enables activities and orients the research effort toward the execution and effects of various scientific projects. As Kuhn puts it (1970, p. 19), "[w]hen the individual scientist can take a paradigm for granted, he need no longer, in his major works, attempt to build his field anew, starting from first principles and justifying the use of each concept introduced. That can be left to the writer of textbooks. Given a textbook, however, the creative scientist can begin his research where it leaves off and thus concentrate exclusively upon the subtlest and most esoteric aspects of the natural phenomena that concern his group". Yet, on the other hand, the property of the aforesaid paradigm causes the action to be disconnected from the motives which determine and explain the shape or course of scientific activity; the power to judge one's own conduct remains outside the area of direct interest of an unreflective researcher executing research projects. Scientific activity devoid of this critical element is barely different, if at all, from the work of an advisor or analyst.

As mentioned, a paradigm includes a certain specific set of philosophical assumptions. The following sections explain which "areas" of scientific activity these assumptions concern, thus answering the question of what basic "components" a paradigm is built of.

2.3   The Structure of Paradigms

2.3.1   Ontological Assumptions, that is Assumptions Concerning the Nature of Social Reality

Ontology is the basic field of philosophy concentrated around the problem of existence. As Ted Benton and Ian Craib (2010) put it, "ontology" is the answer given to the question of what kinds of things exist in the world. The authors point out that in the history of philosophy, four primary traditions in settling this question can be distinguished: materialism, idealism, dualism, and agnosticism. **Materialists** claim that the world is made entirely of matter and the diversity of the features of material objects, living beings—including people, societies, and other beings may be explained in the categories of lower or higher structural complexity of matter. **Idealists** believe that the ultimate reality is of mental or spiritual nature—all beings are the creations of our own internal thought processes. **Dualism** recognizes the separation of the spirit and body, assuming that two interpenetrating worlds exist. And so human beings appear as combinations of a mechanical body and a spiritual mind or soul. Finally, the **agnostic** tradition assumes that it is impossible to discover the nature of the world existing independently of subjective experience (dualism).

As we can see from the descriptions presented above, the essence of the discussion concerns the issue of the **reality of beings**. In reference to ontological assumptions, according to **realism**, something exists "objectively", that is, independently of the human mind. Therefore, ontological realism recognizes that something real exists, that this something actually does exist. In science, this position assumes that the beings and structures referred to in scientific theories exist in some external space, which makes it possible to study, analyze, and learn about them (Heller 2011). From an extreme perspective, sometimes described as "naive realism", it is assumed that in reality, only beings recognized by science exist; only science presents the objective truth about the world (cf. Heller 2011). Extreme realism toward certain universal beings considers them as independent of individuals (e.g. in agnosticism), while moderate realism—as dependent on them. The opposite of realism is the assumption that there is no external, objectively observable and distinguishable structure of a social world. This world consists only of names, concepts, meanings, and terms created and used by people to describe reality. This is why the only thing we can study is the ideas and creations of our minds, not some independent external beings governed by specific laws. It is not difficult to notice that these assumptions are close to idealists, hence this type of view is called **idealism**.

2.3.2 Epistemological Assumptions, that is Assumptions Concerning the Nature of Scientific Cognition

Epistemology is the field of philosophy dealing with cognition and knowledge. Its name comes from Greek *ēpistemē*, which means "knowledge". In literature, one may also encounter the name **theory of knowledge** (Benton and Craib 2010). The fundamental issues of epistemology include the question of the sources of human cognition, the question about the role of experience in the creation of knowledge, the participation of reason in this phenomenon, the problem of the relationship between knowledge and certainty, and other detailed problems concentrated around the issue of the nature of truth, the nature of experience, and the nature of meaning (Woleński 2004).

In seventeenth-century discussions of philosophy and science, there were two main alternative and mutually exclusive views on the nature of cognition. The first one was characterized by a "rationalist" approach, recognizing the primacy of human reason. Its supporters were under the strong influence of mathematics, in which absolutely certain conclusions were reached by way of formal reasoning. The famous Cartesian "Cogito ergo sum" is an example of a statement which, as a result of pure work of the mind—systematically doubting everything—led its author to a reconstruction of the entire edifice of knowledge. The alternative (so-called auxiliary) model was "empiricism". Representatives of this current saw the only source of knowledge about the world in sensory experience. At the moment of birth, the human mind is a blank slate which is filled with experiences during life. This "filling" consists in recognizing and accepting repeatable patterns to which some general ideas are attached. Knowledge is nothing other than isolating these patterns in human experience, and the conclusions drawn from them (Benton and Craib 2010). As eminent experts on the subject claim, currently, a turn in epistemology toward empiricism, that is, a departure from its purely rational version, can be observed (cf.; Hetmański 2008; Sintonen et al. 2004). It is the most intuitive approach to cognition, closest to the common sense of most people. To see, to touch, to smell—means to believe.

It can be concluded that the essence of the debate at the level of epistemological assumptions lies in invoking a certain opposition concerning orientation toward explaining and predicting events in the social world by searching for regularities, laws, and permanent relationships between its elements (variables) (Burrell and Morgan 1985). One of the views is the adoption of a traditional approach of natural sciences in the domain of the social sciences. Making this choice, it should be assumed that knowledge grows in a cumulative manner, and the process consists in adding successive portions of certain and reliable knowledge by way of verification or falsification of new scientific hypotheses. From an extremely opposite point of view, there is no such thing as generally applicable laws regulating the course of social events. Social reality is in its essence relativist and may be explored only in a limited way by attempting to understand it from the point of view of the individuals who experience it. In accordance with this assumption, one should reject the possibility of objective cognition of the social world and adopt the thesis about the multitude of social realities (cf. Schütz 1972), which may be understood only by stepping into the shoes of a participant of the events under analysis. Science is not able to generate any objective knowledge and it is naturally subjectivist.

One of the more interesting ideas making it possible to understand and show the practical application of epistemological assumptions is the concept of **epistemological metaphor** used by the recognized theoretician of organization and management, Gareth Morgan (1980, 1981, 1983, 2006). As Monika Kostera (1996) puts it, Morgan treats research as a kind of involvement in the world, and this is why he proposes setting scientific methods in a broader philosophical context. Metaphor is the expression of the epistemological position of a person designing and conducting research, according to which the ways of perceiving, explaining, and understanding the world recognized by them are more appropriate and valid than others. According to Morgan (2006), we use metaphor whenever we attempt to understand one element of an experience in terms of another. Thanks to metaphors, people "read" reality, express what would otherwise be difficult or impossible to express. "A memorable metaphor has the power to bring two separate domains into cognitive and emotional relation by using language directly appropriate to the one as a lens for seeing the other" (Black 1962, p. 236, as cited in: McCloskey 1983, p. 503).

If we examine metaphor only on the linguistic level, it is nothing other than an analogy, a figure of speech, based on the association of two phenomena and transferring the name of one phenomenon to the other. However, this aspect of metaphor has little value from the point of view of its use in social research. Much more important consequences emerge when metaphor is understood as an internalized form belonging to the sphere of the human psyche, when it is recognized as the basic structural form of experiencing the world. According to Morgan, it is this aspect of metaphor that is of the greatest significance, as it makes it an instrument with which people experience reality, function in it, try to understand, and describe it. The author clearly points it out, among others in the introduction to his book, *Images of Organisation*. In Morgan's opinion, our theories and explanations of organizational life are based on metaphors that lead us to see and understand organizations in distinctive yet partial ways. Metaphor is often just regarded as a device for embellishing discourse, but its significance is much greater than this. The use of metaphor implies a "way of thinking" and a "way of seeing" that pervade how we understand our world generally.

2.3.3 Methodologies

Methodology is treated here as a set of certain choices concerning the way in which a given phenomenon can be studied. David Silverman (2005) thinks that methodologies may be defined broadly and schematically (e.g. methodologies of qualitative and quantitative research) or narrowly and precisely (e.g. the methodology of grounded theory, case analysis, ethnography). In his opinion, a methodology refers to the choices we make about cases to study, methods of data gathering, forms of data analysis, and so on, in planning and executing a research study. So our methodology defines how one goes about studying any phenomenon.

Methodologies are therefore practical ways of learning about the world using various methods of collecting and analyzing quantitative and/or qualitative data. They result from the assumptions, adopted by the researcher more or less consciously, on the nature of the social world and the concept of exploring it. They are therefore a derivative of philosophical assumptions—ontological and epistemological. If, for example, we assume that social reality is of objective nature and can be explored by way of isolating structures and interrelationships between observable manifestations of the phenomenon under study existing outside, then, in conducting research, we will strive to specify the measures of the given phenomenon (variables), the manner of collecting data about the variables (e.g. in the form of a questionnaire of a postal survey), the scale and scope of our activities (selection of the sampling frame and research sample), and the ways of analyzing the collected quantitative material.

Burrell and Morgan (1985) believe that debates conducted in the area of the methodology of social research oscillate around two alternative models of scientific conduct, being a consequence of the choices made at earlier levels of philosophical assumptions—that is, ontological and epistemological. The authors also point out the relationship between methodology and the objectives of social research, and treating as science either ideographic or nomothetic. The first approach—close to the humanist orientation in social research—is based on the conviction that one can only understand the social world by obtaining first-hand information which the participants of events have. Hence the need to come as close as possible to the subject under investigation, analyze the background, context, and history is emphasized here. In ideographic models, it is important to capture the nature of the given phenomenon and create the most complete possible description of that phenomenon by concentrating on details, properties, characteristics, and an in-depth case analysis. The essence of the **nomothetic** approach is to capture the properties of the phenomenon under research which are common, repeatable, and significant for all the cases analyzed. In this type of explanation, the need to control the research process by following established procedures and techniques is emphasized. It is important to maintain the scientific rigor adapted from the natural sciences, which is intended to guarantee objectivity of judgment. Methodologies used to execute tasks of this type use all kinds of standardized methods and instruments of data collection and analysis (e.g. surveys, interview questionnaires, tests).

What are the methodologies of qualitative research characterized by? Monika Kostera (2007) writes that qualitative research strives to achieve as accurate a description as possible of a fragment of social reality undisturbed by the researcher. It is aimed at *understanding*, enlightening, and potentially extrapolating the results to similar situations. Research of this type makes it possible to build comprehensive knowledge—that is, presenting phenomena in their natural context. Knowledge, resulting from qualitative researchers' investigations, is close to the perspective of social actors, it is an attempt at representing the way in which social actors understand reality, what the motives and manners of their actions are. Qualitative methodologies are inductive by nature; they primarily use an idiographic style of explanation.

Matthew Miles and A. Michael Huberman (1994) created a list of qualitative research characteristics recurring in the literature. Among them, the authors name the following:

- Qualitative research is characterized by intense and long-lasting contact with a "field" or life situation which is usually normal; that is, it is a reflection of people's everyday life.
- Researchers try to present the data on the observations of local social actors in a holistic manner, in a way "from the inside", by understanding and suspending (leaving aside) predetermined opinions and judgments.
- From the very beginning of the research process, researchers use research tools which are not overly standardized, taking the active role of formulating questions or the scope of observations, interviews, and analyses upon themselves.

Among the most common, and at the same time the best described methodologies of qualitative research, the following are listed: various versions of grounded theory, ethnographic research, case analysis, focus studies, action research, discourse analysis, critical studies, feminist studies, and the narrative approach.

2.4   Types of Paradigms in Social Sciences

2.4.1   Two Dimensions and Four Paradigms: The Classification of Gibson Burrell and Gareth Morgan (1985)

One of the better-known classifications of paradigms in social sciences was developed over 30 years ago by Gibson Burrell and Gareth Morgan (1985). Using two dimensions from the analysis of the assumptions underlying various social theories, the authors created a matrix of four key paradigms. The first dimension—subjectivism-objectivism—refers to the discussion on the nature of social sciences. The second one—regulation-radical change—concerns the dispute on the nature of society.

The **subjectivism-objectivism** dimension was distinguished as a result of an analysis of the bases of the main debates conducted in the area of ontological, epistemological, and methodological assumptions. The authors of the discussed concept assumed that the approach to practicing science presented by scientists is a sequence of consistent and internally coherent choices within the indicated levels of philosophical assumptions. Adopting specific assumptions about the social reality leads to favoring an adequate form of the theory of cognition, the concept of man, and the practical path of conducting research.

According to the authors, two alternative approaches are possible in social sciences. The first one—**objectivist**—in its purest form assumes that the nature of reality is hard, verifiable, external, and objective (realism). Scientific cognition consists in the analysis of repeatable patterns and relationships between elements of the phenomena under study (nomothetic science), which are primal in relation to the actions of individuals, groups, and communities (social determinism). In order to discover the universal laws governing the social world, one needs to isolate and measure specific variables and the interrelationships between them, using quantitative (statistical) methods.

The second approach—**subjectivist**—emphasizes the primacy of individual experience in the creation of social reality (idealism). In order to understand people and the way in which they act, one needs to come as close as possible to the subject under study, capture and describe how people create their unique worlds while remaining free and, in principle, the primary creators of social reality. Research here is based on an acceptance of the relativist nature of the world and strives to provide the most complete possible description of the analyzed phenomena in order to explore them in depth (ideographic studies; analytic induction).

The **regulation-radical change** dimension, as we mentioned earlier, refers to the issue of the nature of society. Based on an analysis of sociological theories, Burrell and Morgan introduce their own categories, distinguishing *the sociology of radical change* and *the sociology of regulation*. The former is created on the basis of a conviction that man strives to free himself from structures which limit his development potential. Focusing on the conflict-generating nature of interrelationships in society, its representatives aim to point out and explain the existing differences, divisions, and dominances of certain social layers or classes over others. The sociology of regulation on the other hand includes the works of those theoreticians who, first and foremost, strive to present society in categories of unity and coherence. This branch of sociology aims to answer the question of why societies last as single beings.

Burrell and Morgan claim that their distinction of regulation-radical change may serve as an outline of the analysis of social theories. In their opinion, in connection with the subjectivism-objectivism dimension, it constitutes a powerful tool for the identification and analysis of the assumptions underlying all scientific theories. Juxtaposing both dimensions, the authors distinguished four basic paradigms in social sciences: functionalism, interpretivism, radical structuralism, and radical humanism.

The authors of the concept claim that individual paradigms—in spite of the fact that they come into contact with each other—should be perceived as completely distinct from each other. The basis for this distinctness is the irreconcilable assumptions which underlie them. Burrell and Morgan assume that all theoreticians dealing with social analysis may be placed within the presented paradigms. At the same time, they believe that, as with any map, this one also is a tool which shows the current location of the researcher, but also their previous and future locations.

Moreover, it is possible to take various positions within a single paradigm. Below, we present a short description of the paradigms distinguished by Burrell and Morgan.

Functionalism is an approach which dominated the social sciences for a long time. Its representatives perceive society in categories of integrity, invariability, unity, order, *status quo*, searching for universal laws and the reasons for the coherent nature of social reality. The functionalist paradigm is deeply embedded in positivist literature, recognizing the reality and the objective nature of the social world. Functionalists are pragmatists searching for knowledge which is useful and which may serve the purpose of predicting and controlling social processes by providing practical solutions in response to practical problems. This approach assumes that the social world consists of relatively permanent and concrete, empirically available elements, as well as relationships and structures which may be identified, analyzed, and measured using methods borrowed from the natural sciences. In many functionalist theories, comparisons straight from mechanics or biology are used to describe the phenomena under research. Functionalism is based above all on quantitative methods and analyses.

The **interpretive paradigm** in its essence corresponds to the understanding or humanist approach described above. It is deeply embedded in the subjectivist vision of social nature and the theory of cognition. This perspective is oriented toward understanding reality in the form in which it is perceived by its participants (social actors). It searches for explanations by referring to the consciousness, experience, beliefs, and ideas of people who constantly construct and reconstruct their actions. The social world is treated as a continuously emerging and changing social process created by individuals; it is a creation of human minds, a network of assumptions and intersubjectively shared meanings. Researchers from this current focus on everyday life, trying to understand reality, interpreting the social phenomena which occur around them. The interpretive paradigm draws on the assumptions of the sociology of regulation—contrary to appearances, the issues of conflict, domination, and change are not in the center of interest of the researchers of this trend. Rather, they search for explanations concerning the way in which the world is constructed in the everyday activities of social actors.

Radical structuralism at the level of basic philosophical assumptions refers to objectivism and the sociology of radical change. Representatives of this orientation focus on structural relationships within the objectively available reality. They stress that radical change is an inherent part of the nature and structure of the contemporary world. Hence in the center of interest of radical structuralists lie the issues of power, domination, and deeply embedded internal contradictions and disputes. Researchers' activity is not only oriented toward pure description and cognition of these phenomena, but also plays the role of raising society's awareness of the not always fair interrelationships between various individuals and groups. Therefore, science strives to propose ways for unprivileged individuals and groups to free themselves from the domination of others.

Postmodernism, similar to the interpretive paradigm, is based on the conviction that the social world is not a material and objectively available reality, but a product of human minds. A distinctive feature of this approach is perceiving the world in the categories of invalidating social limitations and moving beyond them. One of the fundamental assumptions underlying this orientation is the view that human consciousness is dominated by an ideological superstructure with which man is constantly interacting, and which constitutes a kind of cognitive wedge placed between him and his true consciousness. The existence of this wedge brings alienation; it causes us to deal with "false consciousness", which makes it impossible for us to find fulfilment. The task of this science is to raise awareness of the existence of these cognitive limitations by exposing the false traps of the collective consciousness and freeing the human mind from them, which is intended to lead to self-fulfilment and development of individuals. Society is therefore seen as oriented against man, and radical humanists aim at searching and communicating the ways in which various limitations are imposed on people.

2.4.2 Five Basic Paradigms: The Concept of Egon Guba and Yvonna Lincoln (2005)

Another example of a classification of paradigms in social sciences, with particular consideration given to qualitative research, is the study of Egon

Guba and Yvonna Lincoln (2005) published in *The SAGE Handbook of Qualitative Research*. It presents the authors' proposal for five basic research orientations—updated compared to the first one dating back to 1994. While strongly encouraging everyone to read the above-mentioned publication, only a short and concise description shall be presented here.

The authors analyzed the axiomatic nature of paradigms, referring to three fundamental levels of philosophical assumptions—ontology, epistemology, and methodology. As a result, they distinguished the following paradigms: positivism, postpositivism, critical theory, constructivism, and the participatory paradigm, added in 2000.

Positivism, which we mentioned before, at the ontological level refers to the realistic concept of social reality. It therefore assumes the existence of an objective world, external to the person conducting research; in it, that person searches for laws, rules, and repeatable patterns of activity isolated from a non-significant context. With reference to epistemological assumptions, positivism assumes dualism and objectivism. It is possible to maintain the attitude of an external observer and thus eliminate the influence of values, opinions, and subjective beliefs. The aim of the research is to explain, predict, and control the social phenomena under study. Knowledge, which grows in a cumulative manner, is gathered by way of verification of hypotheses, establishing facts and laws. The methodology is characterized by an experimental approach, with the use of quantitative methods in order to verify the truth of the judgments given.

Postpositivism is a slightly "weaker" version of positivism. In the area of ontological assumptions, it is characterized by critical realism—recognizing the objective nature of reality while assuming that due to the limitations of human senses, it can only be understood in an imperfect way and somewhat approximately. Postpositivists consider the aim of their inquiries to be prediction and control, yet in the area of epistemology, they use a modified version of dualism/objectivism. They find, above all, that it is impossible to completely eliminate the influence of the researcher on the phenomenon under study, but one should aim to reduce it as much as possible. Conducting research, one should also—in reference to the "critical tradition" of scientific research—realize as accurately as possible in what way the research was conducted by subjecting its results and the manner of arriving at these results to the critical evaluation of the scientific community. In the methodological layer, postpositivism assumes critical pluralism, which says that since the human mind has limitations (critical realism), one should aim to diversify the sources and types of data, using various theories, methods, and researchers. So the point is to make active use of the so-called triangulation. Hence in the methodological layer, it is also acceptable to use qualitative methods, even though the quantitative approach is predominant.

Critical theory also borrows the acceptance of the objectivist vision of the world from the natural sciences, yet in the area of epistemology, it presents a subjectivist position, definitely closer to the reality under study. Historical realism—as this is how the authors call it—is characterized by the assumption that reality is shaped by social, political, cultural, economic, ethnic, and other values. The aim of research here is to explain, but also to raise people's awareness of a certain ideological grounding from which one may free oneself in the direction of "true, non-falsified consciousness". Therefore in this case, the reality of beings and the existence of some objective truth is assumed, while at the same time it is recognized that the sociocultural grounding of human actions plays an important role and, in order to explore it, one should refer to people's experience. While in the case of previous paradigms, the researcher took on the role of an uninvolved person focused on providing information to those responsible for introducing changes, in the area of critical theory the researcher is perceived as a spokesperson and activist transforming data so that it becomes comprehensible to the recipient and presenting the position and context of events of the community under study. The methodology is therefore described as dialogic or dialectical, oriented toward debunking false beliefs by reaching them with the use of qualitative and quantitative research, the knowledge of history, the meaning of values, and knowledge of what incapacitation is and what direction the emancipation, liberation, or rehabilitation of individuals should take.

Constructivism in the ontological layer shares the subjectivist attitude toward reality, accepting relativism, that is, the existence of many locally constructed and reconstructed realities. It assumes the existence of various social worlds, functioning above all in human minds, and not as objectively available, common external structures. Some scholars emphasize, however, that constructivism has many variations, and some of them also function in the realist "camp". It is sufficient to assume that certain reconstructions are of a collective nature and assume the shape of an agreement or consensus, thanks to which something becomes something by virtue of shared meanings and ways of understanding the given phenomenon. At the level of epistemological assumptions, constructivists accept subjectivism. In their opinion, it is impossible to separate the researcher from their beliefs and values. Moreover, as realities exist only in human minds and social worlds keep being constructed and reconstructed, the only way of learning about them is to refer to the subjective experience, opinions, beliefs, and values of their creators. And so the researcher assumes the role of "participant", whom Guba and Lincoln call the "facilitator of multi-voice reconstruction". Methodologies are oriented at interpreting meanings, so they are hermeneutical and dialectical in nature. The point is, on the one hand, to bring out certain individual constructs and subject them to interpretation, and on the other, to compare and contrast individual meanings in order to generate one or more shared constructs. Various data sources are used—mainly qualitative, but also quantitative methods and data. The context of events is important.

The participatory paradigm, or the paradigm based on cooperation, at the ontological level recognizes the subjective-objective nature of reality. As explained by John Heron (1996, p. 11), one of the authors of the approach discussed here, reality is subjective, because it is available only in the form in which the human mind presents it. At the same time, it is objective, because a certain given reality, or as Heron puts it, cosmos, that is, a specific harmonious whole, is available to the human mind. Reality is therefore co-created by the mind and the given cosmos. Epistemology assumes interaction, participation in the exchanges between the knower and the known. The roles are interchangeable—the known is also a knower, for we should remember that we are dealing with the social world, not nature. The nature of knowing is practical, participative. Mutual cognition is partial and open to change. At the methodological level, forms of research which fit in the given reality in the practical, conceptual, empathic, and imaginal sense are adopted. Inquiring requires the ability to recognize and build an intersubjective space grounded in a given cultural context. Research based on cooperation is founded on the use of language located in shared experience. The researcher has to have (and they are educated accordingly) emotional competence and a democratic personality, and be actively involved in the given reality.

2.5   Conclusions: Who Needs Knowledge About Paradigms in Social Sciences?

As Normal Denzin and Yvonna Lincoln (2005) claim, the contemporary researcher cannot afford not to know any of the paradigms and perspectives currently practiced in the social sciences. In their opinion, scholars need to understand the basic ethical, ontological, epistemological, and methodological assumptions of paradigms and be able to enter into a dialogue with them. According to the authors, the differences between paradigms have significant and important implications at the practical, everyday, empirical level. To know the basics of the philosophy of social sciences is therefore a duty of every well-educated person dealing with social research. But is it solely a duty? The authors mention the translatability of theoretical perspectives into practical actions. How should we interpret this?

The answer to this question is given, among others, by Ted Benton and Ian Craib (2010), who discuss the "auxiliary" role of philosophy in social sciences. In their opinion, in the auxiliary model, philosophy should provide guidelines and support to the researchers who study the reality around them. This support may be provided in at least three ways:

* Assuming that in our thinking there is bias, prejudice, and indiscriminate assumptions which constitute an obstacle to the progress of science, the role of philosophy may consist in exposing and criticizing them.

- Philosophy may also outline a map presenting the state of scientific knowledge which will make it possible for specialists in individual domains to work out their position in the field of knowledge.
- Finally, philosophers may use their abilities—above all their expertise in logic and argumentation—to perfect research methodologies and methods.

Philosophy is not just an academic discipline. Every person experiences difficult moments in life, when they ponder over fundamental values and principles that guide their actions. At the same time, each of us deals with philosophy in a sense also when we settle the basic problems of everyday life, for example, analyzing our relationships with people and our influence on others, choosing how we spend our free time, or deciding on a job or other activities.

References

Benton, T., & Craib, I. (2010). *Philosophy of Social Science: The Philosophical Foundations of Social Thought* (2nd ed.). Houndsmill/Basingstoke/New York: Palgrave.

Burrell, G., & Morgan, G. (1985). *Sociological Paradigms and Organisational Analysis: Elements of the Sociology of Corporate Life.* Farnham: Routledge.

Denzin, N. K., & Lincoln, Y. S. (Eds.). (2005). *The SAGE Handbook of Qualitative Research* (3rd ed.). Thousand Oaks: Sage.

Giddens, A. (1993). *New Rules of Sociological Method: A Positive Critique of Interpretative Sociologies.* Stanford: Stanford University Press.

Guba, E. G., & Lincoln, Y. S. (2005). Paradigmatic Controversies, Contradictions, and Emerging Confluences. In N. K. Denzin & Y. S. Lincoln (Eds.), *The SAGE Handbook of Qualitative Research* (3rd ed., pp. 191–215). Thousand Oaks: Sage.

Heller, M. (2011). *Philosophy in Science.* Berlin: Springer.

Heron, J. (1996). *Co-Operative Inquiry: Research into the Human Condition.* London: Sage.

Hetmański, M. (2008). Epistemology—Old Dilemmas and New Perspectives. *Dialogue and Universalism, 18*(7/8), 11–28.

Kostera, M. (1996). *Postmodernizm w zarządzaniu* [Postmodernism in Management]. Warsaw: PWE.

Kostera, M. (2007). *Organisational Ethnography: Methods and Inspirations.* Lund: Studentlitteratur.

Kuhn, T. S. (1970). *The Structure of Scientific Revolutions* (2nd ed., enl). Chicago: University of Chicago Press.

McCloskey, D. (1983). The Rhetoric of Economics. *Journal of Economic Literature, 21*(2), 481–517.

Miles, M. B., & Huberman, A. M. (1994). *Qualitative Data Analysis: An Expanded Sourcebook.* Thousand Oaks: Sage.

Morgan, G. (1980). Paradigms, Metaphors, and Puzzle Solving in Organization Theory. *Administrative Science Quarterly, 25*(4), 605–622.

Morgan, G. (1981). The Schismatic Metaphor and Its Implications for Organizational Analysis. *Organization Studies, 2*(1), 23–44.

Morgan, G. (1983). More on Metaphor: Why We Cannot Control Tropes in Administrative Science. *Administrative Science Quarterly, 28*(4), 601–607.

Morgan, G. (2006). *Images of Organization* (Updated ed.). Thousand Oaks: Sage.

Schutz, A. (1972). *Collected Papers I.* M. Natanson (Ed.) (T. 11). Dordrecht: Springer Netherlands.

Silverman, D. (2005). *Doing Qualitative Research: A Practical Handbook.* London: Sage.

Sintonen, M., Wolenski, J., & Niiniluoto, I. (2004). *Handbook of Epistemology.* Dordrecht: Kluwer Academic Publishers.

Woleński, J. (2004). The History of Epistemology. In I. Niiniluoto, M. Sintonen, & J. Woleński (Eds.), *Handbook of Epistemology* (pp. 3–54). Dordrecht: Springer Netherlands.

3

Grounded Theory

Przemysław Hensel and Beata Glinka

3.1   Introduction[1]

Grounded theory is a strategy for conducting qualitative research; its origins in social sciences date back to the 1960s. It is founded on three principles: First, researchers should embark on fieldwork **without having formulated any hypotheses**. By doing so, they ensure that existing theories do not affect their perception of phenomena encountered during fieldwork. Second, the method requires that researchers continuously compare and contrast pieces of collected empirical material. Through such comparison, a set of **codes** is developed and used subsequently for organizing and interpreting the empirical material, and for singling out the most important **categories** that will serve in formulating a theory about the studied phenomenon. Third, the research process is governed by the principle of **theoretical sampling**. Researchers select individuals

P. Hensel (✉) • B. Glinka
University of Warsaw, Warsaw, Poland

© The Author(s) 2018
M. Ciesielska, D. Jemielniak (eds.), *Qualitative Methodologies in Organization Studies*,
https://doi.org/10.1007/978-3-319-65217-7_3

and groups participating in the study in order to expand the researcher's understanding of the problem rather than to form a representative sample, which is common for more traditional methodologies.

In this chapter, we shall present the genesis of grounded theory, its major principles and areas of application. It shall allow the reader to grasp the essential elements of grounded theory, understand basic differences between grounded theory and other research concepts, understand basic principles of research and become familiar with the most common methods and their application in procedures typical of grounded theory, and finally identify cases that best lend themselves to grounded theory methods.

3.2   Origins of Grounded Theory

Grounded theory dates back to the 1960s, when Barney G. Glaser and Anselm L. Strauss published *The Discovery of Grounded Theory: Strategies for Qualitative Research* (1967). Since then, it has become one of the most influential methodological perspectives in social sciences. Grounded theory was created in opposition to the paradigm of deduction-based research process that had prevailed for decades. The deductive process begins with formulating hypotheses based on the existing theories; the collection of data follows; the process culminates in the verification of hypotheses as they are juxtaposed with the data. This deductive procedure has always had as many advocates as critics. Its proponents claim, inter alia, that more general and broader theories can be developed on the basis of logical arguments formulated on a number of a priori assumptions. Deductive procedures simplify the verification of the theory and its potential modification. Critics argue that this approach inhibits the emergence of new concepts; what is more, in the process of verifying the theory, we adjust the reality and the data to the theory's convention, even if its fundamental role is to reflect the reality, not the other way around. Glaser and Strauss (1967) proposed a method that was alternative to the deductive approach and called it "grounded theory." Grounded theory employs **inductive** methods that have a long tradition in social sciences: as Glaser, one of the originators of grounded theory, recently stated, "GT is simply the discovery of emerging patterns in data" (Walsh et al. 2015, p. 593). Glaser and Strauss's approach is strongly related to **symbolic interactionism**, an approach to social reality that can be traced back to the work of George Herbert Mead. According to Blumer, it is based on the following three premises (Blumer 1969/2007, pp. 5–9). Firstly, human beings act toward objects[2] on the basis of the meaning that these objects have for them. Secondly, the meaning of objects is derived from the social interaction between the individual and its environment. Thirdly, the meaning assigned to objects is subject to changes in the processes of interpretation and interaction—not only do individuals conform to the existing meanings, but they also change them.

Given the above premises, grounded theory seems suitable for the exploration of problems related to the perception of social phenomena rather than for the study of the "objective" reality (Suddaby 2006). The qualitative nature of this method renders it particularly suitable for research projects based on case studies (Eisenhardt 1989).

Glaser and Strauss (1967) emphasize that their strategy of comparative analysis focuses on the generation of theory understood as a process: in this sense, a theory is not a finite, complete product, but evolves and grows (just as its "objects," i.e. social phenomena on which the theory is based, change over time).

Comparative analysis proposed by Glaser and Strauss (1967) may result in the emergence of two types of theories: **substantive** and **formal**. Substantive theory means a theory that pertains to a specific empirical area of sociological research, such as race relations, managerial education or managers' professional role. Formal theory deals with a formal, conceptual area of interest, for example, the phenomenon of socialization, the process of succession, deviation, and so on. Both types of theories can be considered **mid-range theories**. Substantive and formal theories operate at a certain level of generalization and differ in terms of their degree of generalization: they may intertwine within a single study. Numerous theories may apply to a single area which, according to Glaser and Strauss (1967), is a positive phenomenon (unlike in the deductive approach). It is sometimes recommended that unexperienced researchers who carry out projects on the basis of grounded theory for the first time should formulate substantive rather than formal theories (Tan 2010).

Grounded theory proves useful to investigators exploring a wide variety of topics. Its methodology should be particularly valuable to the following groups (Martin and Turner 1986): researchers who carry out pilot studies that precede further research on a larger scale; researchers who carry out a case study, which is expected to provide more than just an impression on the examined object; researchers who enquire into these areas in organization that lend themselves best to exploration using qualitative methods, such as institutional work (Lawrence and Suddaby 2006) and the emergence of proto-institutions (Hensel 2018); and researchers who are interested in a systematic and precise collection of facts for the purposes of high quality organizational research.

It should be noted that in recent years, grounded theory has begun to open up to new possibilities afforded by quantitative methods and several interesting mixed-method studies have been published (Walsh et al. 2015). The outline of grounded theory presented in this chapter may therefore be considered an introduction and a description of one among the numerous trends of grounded theory. We encourage readers to explore its varieties and schools of thought (see References).

3.3   Research Strategy

Grounded theory requires the researcher to observe a set of rules of conduct. First of all, in line with the method, the researcher should approach the subject with an open mind (the approach aptly described by Czarniawska-Joerges (1992) as the anthropologic frame of mind), without attempting to formulate hypotheses during the initial stages of research.

Although the concept of grounded theory is based on the absence of any a priori assumptions about the examined issue, it is evident that no researcher, even a novice, is a tabula rasa and always brings to the research process his or her own viewpoint, beliefs and suppositions, which have been collected throughout his or her life (Suddaby 2006). It is important to be aware of these inescapable determinants that influence the cognitive process and to consider them in order to "put them in brackets" (Ashworth 1996; Fischer 2009). In other words, a researcher should try to understand, on the one hand, the extent to which his or her interpretation reflects the analyzed reality and, on the other hand, the influence of his or her preconceptions, preferences and beliefs.

Research carried out in accordance with the rules of grounded theory comprises the following three stages: data collection, coding and identification of ideas, and formulation of theories. It must, however, be emphasized that the above steps are intertwined. Therefore, already at the stage of data collection relevant codes and ideas may be identified, while at the stage of coding and the identification of ideas, the collection of additional data may prove necessary (see Sect. 3.3.1).

3.3.1 Theoretical Sampling and Theoretical Saturation

In the process of collecting data in accordance with the tenets of grounded theory, the principle of **theoretical sampling** should be applied. In traditional research strategies, the process is based on a representative sample, that is, one constructed in a manner allowing every member of a given population to be represented in the analyzed group. In grounded theory, the logic behind the selection of objects for study is different, as the goal is not to "provide a perfect description of an area, but to develop a theory that accounts for much of the relevant behavior" (Glaser and Strauss 1967, pp. 30). The sample must, therefore, provide maximum diversity in terms of research material; when selecting groups and individuals that will form part of the sample, the researcher must try to "generate, to the fullest extent, as many properties of the categories as possible" (Glaser and Strauss 1967, p. 49). The sequence of steps is also different. In studies carried out in accordance with the deductive strategy the definition of the population is followed by random sampling, and the process is concluded with the examination of the objects from the sample. When grounded theory is applied, the size and makeup of the sample are initially unknown; it is subsequently supplemented with new objects added during the research process. It is based on the following premise: we are only capable of deciding whether more observations are necessary after a certain number of observations and interviews have already been carried out.

Data is collected until the point of **theoretical saturation** is reached, which means that further data collection would not expand our knowledge of the examined phenomenon or contribute to the development of the theory. Some authors specify when theoretical saturation is likely to be reached (i.e. upon having carried out a certain numbers of interviews), but such statements should be regarded with caution. In each case, it is the researcher's task to decide whether theoretical saturation has been reached. This may prove problematic to novice researchers and is, therefore, a matter of concern for many of them. It turns out, however, that in the majority of research projects, the theoretical saturation point is relatively easy to identify. Some researchers decide to carry out a few interviews or observations after the theoretical saturation point has been reached; gathering seemingly redundant evidence provides a specific safeguard for their research.

3.3.2   Data Collection and Taking Notes

The first stage, which comprises the collection of data, observations and taking notes, is laborious and time-consuming. Glaser, one of the founders of grounded theory, claims that "all is data" (1998, p. 8). Data collection usually involves numerous observations and interviews[3] (with note-taking). **Observations** carried out within the framework of grounded theory can take many forms: from simple non-participant observation to participant observation to shadowing (understood as being the "shadow" of the research subject and accompanying him/her in daily activities, for instance following a member of the board at work over a period of several days) (Czarniawska 2007). Each of these methods has its advantages and limitations, and the researcher should be aware of them (they are described in detail in other chapters of this textbook[3]). When conducting interviews, researchers usually have recourse to open anthropological interviews, which are not standardized and do not have any strictly defined or homogenous structure. These types of interviews are helpful, because the researcher does not seek to verify previously formulated hypotheses; on the contrary, the purpose is to learn something, identify new and interesting threads. Interviews ought to be recorded, because the researcher should not have to rely solely on his/her memory and notes. In addition, recording allows the researcher to focus on the conversation and take notes that do not relate directly to the content of the interview (which is being recorded), but, for example, describe the circumstances, the environment, facial expressions of the interlocutor and other factors or elements that may be relevant at the stage of coding and interpretation. Before proceeding to the subsequent stage (coding), interviews are transcribed—manually or with the use of computer applications. Transcription should be as accurate as possible, that is, comprise all statements in their original wording along with pauses, hesitations, laughter and other non-verbal elements.

Data obtained through interviews and observations is complementary; it allows the researcher to depict the researched phenomenon more comprehensively. Furthermore, observations often allow researchers to interpret data collected through interviews, or infuse it with a different meaning. It should be noted, however, that the collection of data may go well beyond those traditional areas and include, for example, visual data. As pointed out by Konecki (2008), visual data can be regarded as a valid and valuable empirical material for the formulation of theories or theoretical proposals.

The basic principle is the **triangulation of data**; its purpose is to guarantee greater credibility of the collected evidence and its interpretation. Denzin (1970/2006) distinguishes four basic types of triangulation:

- **methodological** triangulation—combination of different research methods
- **data** triangulation—using various data (e.g. comparing research carried out among different groups, at different times and in different places)
- **investigator** triangulation—involving multiple researchers (at the stage of collecting material and/or its interpretation)
- **theory** triangulation—use of different theoretical concepts to explain social phenomena

Taking notes is crucial: they are the starting point for the identification of categories, or ideas, that will serve as a basis of theory. Martin and Turner (1986, p. 145) define good notes as follows:

1. They provide complete descriptions of the studied situations, are full of details and tell "stories" about the events.
2. They are more than a chronological record of the situation: they shed light on the context of the events described.
3. Good notes contain as few comments as possible; if comments are necessary, they should be clearly marked as such.

Good notes should remain useful for a long time after the research and they can serve as a source of inspiration for future research projects. In grounded theory, data is collected until the theoretical saturation point is reached, which often entails a long research process, especially in the case of complex problems (see the subsection on Sect. 3.3.1). Grounded theory is therefore not a good option when time is limited. If a student decides to rely on grounded theory to prepare his/her MA thesis with only a few months left before the deadline, it is probably not a good idea. Even if the thesis is submitted on time, it is likely to be incomplete, unfinished or based on insufficient data.

3.3.3   Coding and Identification of Ideas/Concepts

Coding is a phase in the research process during which the researcher's attention shifts from data to abstract categories. Coding is one of the most important stages of a research project carried out in accordance with the principles of grounded theory, as it is during this very phase of the study that specific categories emerge; in the final phase, they will form the basis for the mid-range theory.

Codes should be adjusted to the empirical material and reflect the events observed or the narratives provided by the respondents. Any references to theoretical concepts should be avoided to minimize the risk of data being analyzed through a theoretical lens—after all, the aim of the grounded theory is to proceed in the opposite direction: empirical evidence should lead to development of new theory.

It should be emphasized that qualitative coding used in grounded theory rests on a completely different premise than quantitative coding. In the latter, codes and categories exist prior to the coding process, and they typically originate in existing theories. In qualitative coding, codes "emerge from the field," which means that they are generated during the coding process and derive from it (Charmaz 1983).

We also need to remember that the term "code" is used in a variety of contexts even by the precursors of grounded theory (Locke 2001). Coding can be construed as the process of naming individual fragments of observation, but also as assigning observations to specific categories and the theoretical analysis of the created codes.

Regardless of the adopted coding strategy, the crux of this phase is the **constant comparison** of the various elements that form part of the collected empirical material. It results in the creation of interesting codes that, in turn, allow the researcher to formulate compelling conceptualizations and captivating theories.

Clearly, the meaning of codes is determined by the **context of comparisons**. A quote becomes meaningful only after it is juxtaposed with other statements on a similar topic. Codes and categories are created as a result of comparing statements; it also provides them with a specific meaning.

A variety of **coding strategies** are applied to empirical material (Charmaz 2006, pp. 42–71): word-by-word coding, line-by-line coding or incident-to-incident coding. The three coding strategies have the same goal, namely, they let the researcher look at the empirical evidence with the anthropologic frame of mind, in order to perceive new phenomena within well-known and seemingly trivial behaviors and descriptions of the world. Through word-by-word coding, the researcher focuses on nuances, on the manner in which respondents express their thoughts and on details that might elude him/her if other methods of coding were applied. Line-by-line coding lets us look at the coded material through the prism of the division imposed by the width of a column of text, which rarely coincides with the logic of the analyzed material. In the case of incident-to-incident coding, text analysis echoes to the greatest extent our natural perception of the narrative. This type of coding emphasizes the chronological order and reveals the sequence of events, along with the broader context in which they occur. This coding method is also closest to the original proposal advanced by Glaser and Strauss.

The choice of the coding strategy depends on many factors, including the length of the text itself. Word-by-word coding is exceptionally well suited to the analysis of short documents from the studied organization; nevertheless, applying this method of coding to several hours of interviews would be too time-consuming. Hence, any potential benefits would be lost given the effort expanded.

A distinct coding strategy is the so-called **in vivo coding** (i.e. "live coding") (Charmaz 2006, pp. 55–57). This approach involves the creation of codes based on terms commonly employed within the examined organization: terms that all members of the organization are familiar with. In other words, instead of being developed in the language of the researcher, codes are based on the language of respondents.

When specific phrases used within the organization are used as codes, the researcher can detect different meanings attributed to these phrases; this, in turn, may reveal certain interesting dimensions of the organizational reality. In vivo codes are, in general, specific terms used to denote human types, objects, organizational and technological solutions. Researchers should be aware of in vivo coding already at the stage of interviews and observations. They must not assume that the respondent and the researcher understand commonly used terms in the same manner.

Coded descriptions are grouped into **categories** and labelled/named as specific phenomena. The aim of this stage is not to organize the notes, but to replace thinking about specific incidents with a more abstract analysis. Grouping can be done, for example, on the margins of notes (transcripts of interviews) or—as proposed by Martin and Turner—using the so-called **concepts cards**. A particular description can be assigned to different categories, labels (or written on different cards), especially during initial research phases. Nowadays, coding is usually conducted with the use of dedicated software packages, such as Atlast.ti or NViVo. The longer the coding process, the less emphasis is placed on comparing new data with other data within the same category. The focus is shifted onto comparing new data with the idea represented by this category. In other words, we become less interested in comparing a new citation—concerning, for instance, the perception of organizational structures—to previously analyzed citations; instead, we strive to understand how it contributes to our understanding of the category that has been labelled the "perception of organizational structure."

We present below an example of the use of in vivo coding and the creation of concept cards in the study of the language employed by CEOs of large Polish companies operating in the FMCG sector in the context of discussing organizational changes.

Example 3.1 Examples of Results Obtained with Grounded Theory Methods. Source: Authors' Own

Czarniawska-Joerges (1988) identifies four main linguistic tools that are applied to construct the organizational reality: labels, metaphors, clichés and irony. Labels serve primarily the purpose of classification. For a phrase or expression to become a label, it must be repeated time and time again. It is through repetition and dissemination that a phrase gains its categorizing power. Labels are essential in the process of constructing reality, primarily because prior to measuring up with a problem, we have to name it. Naming the problem presents multiple advantages: first and foremost, what is named becomes familiar. Once a phenomenon is named, we feel more assured. The same mechanism is observed, for instance, when members of aboriginal tribes name meteorological phenomena after deities, in an attempt to "tame" them. Labelling allows us to structure the world around us: through naming specific phenomena, we define relationships between them.

Czarniawska-Joerges believes that labelling is a powerful tool of influence within the organization. More often than not, the name we attribute to a phenomenon contains a value-laden element and indications about the future (i.e. what should be done with this phenomenon). Those who are able to convince others to adopt their labels hold a mighty tool with which they can exercise power. Weick (1985) points to the role of labels in organizational life:

> Labels carry their own implications for action, and that is why they are so successful in the management of ambiguity. Consider these labels: that is a cost (minimize it), that is a spoilage (reduce it) … this is a stupidity (exploit it), and so forth. In each of these instances a label consolidates bits and pieces of data, gives the meaning, suggests appropriate action, implies a diagnosis, and removes ambiguity. (Weick 1985, p. 128, quoted in Czarniawska-Joerges 1988, p. 15)

Types of labels depend on the particular situation of a given organization. Labels in times of change differ from those that are used when the organization does not go through any spectacular transformation. What is more, a label may denote different phenomena and its value is contingent on the current situation of the organization.
(…)

Through coding and analysis of the material, a number of labels used by the managers of the analyzed company have been distinguished. By far, the strongest label is **"structure."**

Extracts from the concept card for the "structure" code
"I have been convinced from the very beginning that these problems are due to the organizational structure."
"This new structure is completely different from the previous one. You never step into the same river twice."
"I have had here my fair share of failures, and this is due to the structure of the company: it is not open to outsiders."

The label was mentioned several times during each interview with members of the Board. Different meanings were assigned to the "structure" label: some defined it as the division of tasks and responsibilities within the organization, others understood it as a hierarchy or nomenclature that does not accept outsiders.

It is linked to a number of secondary labels: centralization, decentralization and divisionalization.

Extracts from the "centralization" code concept card
"Prior to these changes, the company was managed in a rather centralist manner."
"The abolition of centralism meant that the staff very quickly felt there was no one supervising them, which immediately brought disastrous effects."
"In Communist times, the enterprise was centralized to the same extent everything was centralized during that period."
"Later, during the transformation period, some things could not be centrally controlled and certain decisions were made, granting a high degree of autonomy to these entities."

Extracts from the "decentralization" code concept card
"A decision was made to maximize decentralization."
"To this day, this maximum decentralization is in progress."
"After a period of overall decentralization, we went back to centralization and the process of renewed, reasonable centralization is still ongoing."
"I tried to get what I could out of that decentralized structure."

Extracts from the "divisionalization" code concept card
"The new divisional structure will allow us to organize everything."
"Due to the resistance of factory floor staff, the process of divisionalisation is not, in fact, what we would like it to be."
"We are now starting to create a divisional structure."

In order to explain the importance attached by respondents to the structure label, we must look back into the company's history. In recent years, the company has undergone a thorough restructuring process with three distinct stages. At the beginning, it was managed centrally, from its headquarters in Warsaw. In 1989, the company initiated a far-reaching decentralization process, only to eventually revert to the centralization of management several years later.

The consequence of the above is a certain confusion related to the valuation of labels. Labels that once were considered positive (e.g. decentralization), now have a negative status, because the management believes that the majority of the company's problems have their roots in excessive decentralization. The opposite has been observed in the case of the centralization label: once a symbol of mismanagement and poor structure, it is now promoted as a remedy.

Managers do their best to make sure that labels are understood in the way they "should" be understood at a given point in time. Thus, they are almost always qualified with adjectives that help distinguish the old meaning from the new.

Table 1 Adjectives affecting the valuation of labels

Label	Positive qualification	Negative qualification
Structure	New	Old
Centralization	Reasonable, certain, new, similar to the centralization of international corporations	Communist-style
Decentralization	Reasonable	Maximum, wide-ranging
Divisionalization	Allowing to organize everything	

Divisionalization is a particular example, as this label has replaced the previous centralization label. Therefore, instead of saying, "we revert to centralization," the company refers to "divisionalization" which, in the case of this entity, is synonymous with centralization.

Codes and categories are not rated as more or less true, but as more or less useful. Over time, irrelevant codes cease to be used, whereas useful concepts remain helpful. Through the analysis of relations between the most useful categories, the **core category** is formulated. It then becomes the most important axis of the theory that is being created. The analysis that allows the researcher to explore the relationship between codes and concepts is called axial coding. When selecting categories, our primary goal should be to ensure that the level of abstraction is high enough to avoid creating a separate category for each incident, yet sufficiently low to ensure that each concept relates to a specific phenomenon. It should be emphasized that little is irreversible in grounded theory: if categories prove excessively detailed or abstract, they can be subsequently modified. The process is, by definition, non-linear and iterative, and therefore nothing prevents the researcher from going back to the previous phase and, for instance, creating a new category at any stage of the research process. The primary and over-arching methodological requirement is fidelity toward the empirical data.

Glaser (1978) distinguishes two types of coding: substantive and theoretical. In his view, substantive codes conceptualize the empirical substance of the researched area, while theoretical codes conceptualize the manner in which substantive codes may relate to each other. In other words, theoretical coding allows us to integrate substantive codes (Hernandez 2009; Holton 2010). Glaser also proposes "a list of 'coding families' that contain codes unified through logical and formal relationships (e.g. elements, divisions, characteristics, sectors, segments etc.) or the dimension of relationships (e.g. strategies, tactics, achievements, manipulation, maneuvering, intrigues, meaning, goals, etc.)" (Konecki 2008, pp. 91).

He proposes a total of 18 coding families (which are not mutually exclusive), including, for example, the "The Six Cs" family (causes, contexts, contingencies, consequences, covariances and conditions), the "Process" family (including stages, phases, transitions, passages, careers, chains and sequences), or the "Degree" family (comprising, inter alia, extent, level, intensity, range and critical points) (Konecki 2008, pp. 91–92). Other approaches to coding can also be found in the extant literature (e.g. Strauss and Corbin 1990).

3.3.4 Generating Theory

It is, by definition, impossible to formulate a single recipe for developing a valid scientific theory. This creative process is complex and the application of an imposed algorithm does not guarantee good results (Strauss and Corbin 1990; Suddaby 2006).

The crux of this concept—understood as a formula for developing a theory—lies in the integration of the three phases of data collection mentioned above. As the authors of the concept argue, "Joint collection, coding, and analysis of data is the underlying operation. The generation of theory, coupled with the notion of theory as process, requires that all three operations be done together as much as possible" (Glaser and Strauss 1967, p. 43).

As the research process unfolds, the investigator better understands the studied phenomena; the codes, categories and hypotheses formulated begin to reflect reality to a greater extent, creating a grid of meanings around the core category. The core category should play a central role and explain to a great extent the described behavior, refer to many observations and be significantly correlated with other categories (Goulding 2002).

It must, however, be emphasized that theory does not emerge from the data itself; data does not "choose" the manner in which it wishes to be recounted (Locke 2001, p. 53). In fact, the researcher keeps making decisions that shape to a great extent the theory that will be formulated at the end of the research process. The researcher decides which data will be taken into account and which will be discarded, what seems interesting and what is banal, and so on.

According to Glaser and Strauss, a theory consists of **categories** and **properties**, as well as the relationships between them. Both derive from the data that has been collected and analyzed. What is the difference between them? An example of a category is the "perception of modern management techniques," while a property is the justification of the use of such techniques. Relations between categories are discovered through comparison. At this stage, it is recommended to take notes and draw diagrams to visualize such comparisons (Locke 2001, p. 52).

One of the stages of generating a theory is the formulation of hypotheses that emerge as a result of comparing the empirical evidence collected in different places or derived from interviews with various interlocutors. Glaser and Strauss emphasize that what we deal with throughout the process is the logic of exploring the theory, and therefore we do not need to concern ourselves with the number of cases on the basis of which we hypothesize: the aim is not to verify on the basis of a representative sample, but to develop a theory.

In the following step, hypotheses are integrated: thus, the theory is developed. An integrated theory emerges from the data, and therefore the researcher should not be tempted to try and match the theory with other concepts derived from the extant literature. Literature research can only supplement the collected data and it should not have any substantial impact on the final theory.

At this stage, the researcher can also note references to literature and existing theories (both Glaser (1978) and Turner (1981) object to referring to literature at earlier stages in order to minimize the impact of existing theories on the process of formulating one's own concepts related to the subject of research; nevertheless, Strauss and Corbin recommend tapping into the existing sources during the early stages of the research[4]). The process of in-depth defining of categories is non-linear and iterative: certain common threads may start to appear at an early stage of research, and one often goes back to previously described phenomena.

3.4   Misconceptions Related to Grounded Theory

The grounded theory method is based on premises that strongly differ from those typical of traditional, deductive research (see Sect. 3.1). At the same time, it is an iterative method, which requires the investigator to repeatedly go back to earlier stages of the research. Conflicts around the "right" understanding of grounded theory are all the more understandable when we take into account that even the authors of the original concept eventually "fell out" and each of them advanced his own, distinct version of the theory. In his later studies, Glaser emphasizes the emergence of theory from data, while Strauss (in cooperation with Corbin) advocates the use of strict coding rules (Goulding 2002, pp. 158–160). Glaser argues that Strauss's suggestion contradicts the very premise of grounded theory (Glaser 1992). Glaser's approach seems to put a greater emphasis on creativity, potentially at the expense of theoretical cohesion, while the approach represented by Strauss and Corbin poses the risk of excessive formalism and inflexibility (Fendt and Sachs 2008). The researcher must decide on one of the two versions of the theory prior to the commencement of research, as striving to combine them may prove unsuccessful.

Given the above, scholars who decide to use the method for the first time may feel confused. Let us, therefore, clarify what grounded theory **is not** (Suddaby 2006):

(a) **Grounded theory is not an excuse to ignore literature**. As noted in the Introduction, a researcher is not a tabula rasa, has his/her own thoughts, opinions and at least partial knowledge of literature. No researcher embarks on fieldwork without having formulated the problem he/she intends to examine. A general idea about the topic of the study is necessary: without it, the researcher is likely to collect a wealth of empirical evidence that will prove impossible to organize and will yield no conclusions. Literature can be instrumental in defining crucial categories that will help analyze the researched reality. Detachment from the existing research, suggested by Glaser and Strauss, means first and foremost that one should not regard the findings of other researchers as absolute and unquestionable, but instead look for new categories, concepts and codes where it is possible and justified.

(b) **Grounded theory is not presentation of the raw data**. According to Suddaby, many reports from studies carried out in accordance with the methodology of grounded theory do not have any substantial cognitive value because, in fact, they do not present a theory, but only a description of events that took place within the observed social reality. In other words, some authors fail to move up to a higher level of abstraction: instead of exposing relationship between categories and their properties, they present incidents organized in categories.

(c) **Grounded theory is not theory testing, content analysis, or word counts**. As noted earlier, the method in question is expected to result in the development of a new theory and not in verifying existing hypotheses. For that reason alone, it makes no sense to use grounded theory to test hypotheses. In addition, there is a great risk that in the process of testing hypotheses, the researcher will identify within the collected evidence only the categories that are related to the previously formulated hypothesis, and not those stemming from the empirical data.

(d) **Grounded theory is not simply routine application of formulaic technique to data.** As mentioned in the section on theory generation, this approach cannot be understood as a simple recipe for good research results. Although a number of general recommendations about methods of conducting research in accordance with the tenets of grounded theory can be formulated, in no way does their observance guarantee the development of a new and valid theory. Neither a strict application of coding guidelines nor the use of modern software packages (e.g. Atlas.ti or NVivo) to analyze the empirical material can replace the researcher's skills of reading and interpreting the collected evidence.

(e) **Grounded theory is not perfect.** According to Suddaby, the great popularity of this method has attracted a number of scholars who focus only on its theoretical aspects, without ever undertaking any empirical studies. They are responsible for promoting the dogmatic version of grounded theory and recommend to future researchers strict adherence to their guidelines that are supposed to facilitate research (e.g. one of these tips is that theoretical saturation is usually reached after 25–30 interviews). Tensions between theorists and practitioners are a natural phenomenon, yet we need to keep in mind that the techniques of grounded theory are inherently a little "messy" (Suddaby 2006, p. 638); the primary goal should always be the development of a valid theory.

(f) **Grounded theory is not easy.** Contrary to what one might think after having read the best reports presenting the results of research conducted in accordance with the tenets of grounded theory, it is not an easy research strategy. Not only does it require the understanding of the method itself, but also the ability to discern the impact of researchers' beliefs and experiences on the research process (see Sect. 3.3). What is more, not all scholars possess the same interpretation skills and the same ability to discern patterns within the collected material.

Notes

1. The chapter contains extracts from Przemysław Hensel's unpublished MA thesis entitled *Język podczas zmiany kulturowej* (Language in the process of cultural change), Faculty of Management, University of Warsaw, 1997.
2. For Blumer, the word "object" has a broad meaning: it encompasses physical artifacts, abstract concepts and institutions.
3. Both interviews and observations have been extensively discussed in another chapter of this textbook.
4. For a discussion on the pros and cons of the two approaches to literature review, see McGhee et al. (2007).

References

Ashworth, P. (1996). Presuppose Nothing! The Suspension of Assumptions in Phenomenological Psychological Methodology. *Journal of Phenomenological Psychology, 27*(1), 1–25.

Blumer, H. (1969). *Symbolic Interactionism: Perspective and Method*. Englewood Cliffs: Prentice-Hall.

Charmaz, K. (1983). The Grounded Theory Method: An Explication and Interpretation. In R. Emerson (Ed.), *Contemporary Field Research: A Collection of Readings*. Boston: Little Brown.

Charmaz, K. (2006). *Constructing Grounded Theory. A Practical Guide through Qualitative Analysis*. London: Sage.

Czarniawska, B. (2007). *Shadowing: And Other Techniques for Doing Fieldwork in Modern Societies*. Copenhagen: Copenhagen Business School Press.

Czarniawska-Joerges, B. (1988). *To Coin a Phrase: The Study of Power and Democracy in Sweden*. Stockholm: Regeringskansliets offsetcentral.

Czarniawska-Joerges, B. (1992). *Exploring Complex Organizations: A Cultural Perspective*. Newbury Park/London/New Delhi: Sage.

Denzin, N. K. (1970/2006). *Sociological Methods: A Sourcebook*. New Brunswick: Transaction Publishers.

Eisenhardt, K. M. (1989). Building Theories from Case Study Research. *Academy of Management Review, 14*(4), 532–550.

Fendt, J., & Sachs, W. (2008). Grounded Theory Method in Management Research: Users' Perspectives. *Organizational Research Methods, 11*(3), 430–455.

Fischer, C. T. (2009). Bracketing in Qualitative Research: Conceptual and Practical Matters. *Psychotherapy Research, 19*(4/5), 583–590.

Glaser, B. G. (1978). *Theoretical Sensitivity: Advances In the Methodology of Grounded Theory.* Mill Valley: Sociology Press.

Glaser, B. G. (1992). *Basics of Grounded Theory Analysis: Emergence vs Forcing.* Mill Valley: Sociology Press.

Glaser, B. G. (1998). *Doing Grounded Theory: Issues and Discussions.* Mill Valley: Sociology Press.

Glaser, B. G., & Strauss, A. L. (1967). *The Discovery of Grounded Theory: Strategies for Qualitative Research.* Hawthorne: Aldine De Gruyter.

Goulding, C. (2002). *Grounded Theory: A Practical Guide for Management, Business and Market Researchers.* London: Sage.

Hensel, P. (2018). Organizational Responses to Proto-Institutions. *Journal of Management Inquiry, 27*(2).

Hernandez, C. A. (2009). Theoretical Coding in Grounded Theory Methodology. *The Grounded Theory Review, 8*(3), 51–60.

Holton, J. A. (2010). The Coding Process and Its Challenges. *The Grounded Theory Review, 9*(1), 21–38.

Konecki, K. (2008). Wizualna teoria ugruntowana. Rodziny kodowania wykorzystywane w analizie wizualnej. *Przegląd Socjologii Jakościowej, 3*, 89–115.

Lawrence, T. B., & Suddaby, R. (2006). Institutions and Institutional Work. In S. R. Clegg, C. Hardy, T. B. Lawrence, & W. R. Nord (Eds.), *Handbook of Organization Studies* (pp. 215–254). London: Sage.

Locke, K. (2001). *Grounded Theory in Management Research.* London: Sage.

Martin, P. Y., & Turner, B. A. (1986). Grounded Theory and Organizational Research. *The Journal of Applied Behavioral Science, 22*(2), 141–157.

McGhee, G., Marland, G. R., & Atkinson, J. (2007). Grounded Theory Research: Literature Reviewing and Reflexivity. *Journal of Advanced Nursing, 60*(3), 334–342.

Strauss, A., & Corbin, J. (1990). *Basics of Qualitative Research.* London: Sage.

Suddaby, R. (2006). From the Editors: What Grounded Theory Is Not. *Academy of Management Journal, 49*(4), 633–642.

Tan, J. (2010). Grounded Theory in Practice: Issues and Discussion for New Qualitative Researchers. *Journal of Documentation, 66*(1), 93–112.

Turner, B. A. (1981). Some Practical Aspects of Qualitative Data Analysis: One Way of Organizing the Cognitive Processes Associated with the Generation of Grounded Theory. *Quality and Quantity, 15*, 225–224.

Walsh, I., Holton, J. A., Bailyn, L., Fernandez, W., Levina, N., & Glaser, B. (2015). What Grounded Theory Is … A Critically Reflective Conversation Among Scholars. *Organizational Research Methods, 18*(4), 581–599.

Weick, K. (1985). Sources of Order in Underoganized Systems: Themes in Recent Organizational Theory. In Y. S. Lincoln (Ed.), *Organizational Theory and Inquiry. The Paradigm Revolution* (pp. 106–136). Sage: Beverly Hills/London/New Delhi.

4

Visual Anthropology

Slawomir Magala

4.1 Introduction

The aim of this chapter is to explain what visual anthropology is and to demonstrate her (and her sister's, visual sociology's) uses in understanding the increasingly sophisticated competition for our (un)divided attention. This fierce, mostly visual and iconic, competition underpins systematically mobilized and individualized interactions and communications. These intense, passionate, individualized, mobile communications happen under our very eyes—otherwise known as processes, ebbs and flows of social life, they accelerate with the mobile multimedia connectivity.

Navigating our increasingly complex social worlds (real, fictitious and virtual), comprehending these increasingly sophisticated cultural realities (real, virtual and interactive), requires skills, access to sociocultural capital

S. Magala (✉)
Erasmus University, Rotterdam, The Netherlands

Jagiellonian University, Kraków, Poland

© The Author(s) 2018
M. Ciesielska, D. Jemielniak (eds.), *Qualitative Methodologies in Organization Studies*,
https://doi.org/10.1007/978-3-319-65217-7_4

and empowerment to employ the above-mentioned resources. Hence the growth of interest in visual communications, which facilitate (what do they facilitate?—learning outside of the range of control status gatekeepers), accelerate (what do they accelerate?—response, which can be individually articulated outside of institutional frames) and equalize (what do they equalize?—with the spread of higher education and mobile phone access to individualized information, knowledge and competence have been driven down the social ladders). Growth of research interest, but also growth of ideological and political hopes ("emancipation begins when we challenge the opposition between viewing and acting" Ranciere 2009, p. 13), are among the most obvious results. Icons are not innocent, nor is a tracing eye impartial (Goffman 1988).

4.2   What Is "Visual Anthropology"?

Visual anthropology and sociology of visual culture are the closest outposts of academically legitimized research community in these newly developing, emergent, contingent territories, marked as "white spots", "black holes" or "terra incognita" of visual communications on our "knowledge" maps produced through the academic division of labor. Matters of visual communications challenge our cognitive, academically trained talents—most of us had been raised, brought up and trained in and on books (i.e. on records, on written documents produced either as renderings of oral lectures or already from the very beginning designed and composed as written reports from earlier research activities). These books have mostly been written by older white males, whose dominant position is nowadays being questioned, not in the least by younger, more female and less white experts in visual communications. Visual anthropology and visual sociology are not brand new appearances on an academic stage. Many researchers had signaled interest in how to draw and maintain attention in "global villages" (cf. McLuhan 1964), how to resist hidden persuaders working through "all consuming images" (Ewen 1988), how to explain skills in attracting audiences and spectators within a postulated political economy of attention (Franck 1998, 2005), how to view images as a new "alphabet" of communication (Mitchell 2005) and how to develop competence in communicatively hyperactive "network society" (Castells 1996, 2009). All these studies increase our knowledge of visual communications, improve our competence in negotiating through interactions and communications and empower us as individuals and citizens to implement our newly won knowledge and competence in "making our way through the world" (Archer 2007, Hindletter et al. 2009). These efforts are slowly maturing and the latest developments in visual anthropology/sociology and sociocultural mobilization (even if only as "flash mobs" in shopping malls) are the case in point.

The terms "visual anthropology" and "visual sociology" are used interchangeably—with "visual anthropology" in the lead—due to the fact that cultural anthropologists investigating distant, "exotic" cultures and communities had started to use photographic reports before sociologists followed them in asking photographers to document visually "how the other half lives" (Jacob Riis in New York City) or how the impoverished tenant farmers leave the dust bowl of Midwest, trekking toward California (Roy Stryker and Dorothea Lange for Farm Security Administration under the New Deal policies implemented by Franklin Delano Roosevelt). At present, the *anthropology* of a visual image is a fast-growing domain of knowledge, but so is the *sociology* of visual representation, of multimedia and of visual culture (cf. Worth 1981; Levin 1993; Anker and Nelkin 2003; Sikora 2004; Olechnicki 2003a, b; Dikovitskaya 2005; Drozdowski and Krajewski 2010). The reason for this overlap is to be found in a profoundly hybrid composition of major theoretical inputs, which generated a theoretical frame for both visual anthropology and sociology of visual culture. These inputs include the semiology, sociology, aesthetics, ethics and politics of photography and of the arts inspired or influenced by photography and mechanical reproduction of images (this line of research and criticism has been kindled by Roland Barthes and Susan Sontag, cf. Barthes 1980; Sontag 1977, 2001, 2003; Crimp 1997, Butler 2006), answering the questions: "every picture tells a story: or doesn't it?" or "Photography is not an opinion. Or is it?" (Sontag 2001; Hariman and Lucaites 2007). Further, these inputs include the aesthetics of multimediated artistic communication (Panofsky 1982; Duve 1998; Belting 2003; Foster 2004; Mitchell 2005) and from media lab studies of the MIT critics and researchers often associated with Rosalind Kraus and the "October" quarterly, which she had co-founded with Anette Michelson (the editorial board includes Yve-Alain Bois, Benjamin Buchloh and Hal Foster, household names in contemporary art and multimedia criticism).

The emergent, though mostly tacit, consensus among researchers in visual studies, no matter whether they tend to label themselves as "visual anthropologists" or "visual sociologists", is that anthropology/sociology of images should be an interdisciplinary domain *per se* and that its location on the "map" of human knowledge should situate it in a triangle between:

- **social sciences and the humanities** trying to increase our understanding of sociocultural flows of interactions, communications and sophisticated feedbacks between them (a deliverable product of visual anthropological/sociological research should be a body of collectively accessible scientific/scholarly knowledge—Pauwels or Castells are cases in point; cf. Pauwels 2006, 2009; Castells 2001, 2009).
- **contemporary art and artistic practices** underlying/undermining communications and trying to improve our sensemaking practices through the expansion of cultural resources (a deliverable product of visual anthropological/sociological explanation and interpretation should be an improved array of skills at selective, sophisticated, competent sensemaking—e.g. Wells 1997; Manovich 2001; Elkins 2007).
- **"daily life"** as an increasingly complex and unpredictable flow of interactions and communications (a deliverable product of visual anthropology/sociology should be the "added value" of individually acquired, critically examined, meaningful experience—e.g. Turkle 1995; Harriman and Lucaites 2007; Costello and Willsdon 2008; Wasik 2009).

4.3   Photography: The Founding Mother of Visual Anthropology/Sociology

Let us begin with the respectable, legitimate knowledge, as it is built by visual communications of research—very early on it has been noted that a photograph can be viewed as a "record" about culture (i.e. as a visual document reporting on a cultural reality under consideration) or as a record "of" culture (i.e. as a visual artifact created within a framework of a meaningful activity typical for a given community—for instance, when a proud father takes a picture of his son or daughter receiving the first holy communion, or a moved mother photographs her son getting married or when both produce images of their offspring graduating from a school in order to frame them and hang them in their sitting rooms). Sol Worth spoke of a photograph "about" culture taken from the outside, and a photograph symbolically representing this culture, taken by the "insider". Sikora speaks of a photograph functioning as a document ("about") and as a symbol ("of"). Barthes spoke of the "stadium" of obvious topic of an image and of the "punctum" of a photograph, which "pierces" the viewer with an illumination of a symbolic "message" (in his brilliant essay "Camera Lucida", whose title is an obvious pun on "Camera Obscura").

Roughly speaking, the documentary function, while still very significant, slowly and gradually fades away as more sophisticated, skillfully composed and culturally "tainted" uses of photograph and other forms of communicated image are filling our communicative spaces. The classical approach is well represented by Piotr Sztompka, former president of the International Sociological Association, who acknowledges the "visual turn" in contemporary social sciences in general, and in visual anthropology and sociology in particular (his latest book is even entitled—quite simply—*Visual Sociology*, cf. Sztompka 2005). Sztompka sees photographs as documentary evidence, on a par with the less visual, more analytical and verbally executed research procedures and methodologies. He mentions the late Polish photographer, Zofia Rydet, who had become famous with a huge project entitled "Sociological Record", executed in the years 1978–1990. "Sociological Record" involved hundreds of "posed" portraits of older, poor or marginalized individuals in small towns or villages of the traditional industrial region of Silesia undergoing the dramatic transformations after the decline of the steel industry and coal-mining, which had once made it very prosperous. The problem, however, is that Sztompka believes Rydet prima facie takes her word for granted, when she talks about "sociological record"—he thinks that she means recorded documentary evidence of empirical "sociological investigation", or analytical "sociological research" *tout court*. However, this is not the case. Her photographs are as far away from a documentary report on "real life" of the poor and the marginalized individuals in Poland of the 1970s and 1980s as are August Sander's photographs of Germans of all ranks and files, posing for him in their best clothes or in their professional uniforms, demonstrating their identities, their pride in their social status and professional recognition in the 1920s.

Neither Rydet nor Sander left us with innocent anthropological or sociological records, pure documents of objective research activities. Their photographs are profoundly symbolic—they were made by conscious and competent artists exactly the way the depicted members of local communities would have liked to be seen from the outside (which they made clear to their sensitive, gifted and understanding "registrators"). They—Sander and Rydet—took the ethnomethodological approach seriously (though I doubt if they ever heard of the "theatre of daily life" and "social construction of reality" inspired by phenomenologists) because this seemed to them to fit most closely with their artistic project of symbolic representation of a society (Germans after WWI in case of Sander) or of a region (Silesian provincial towns and villages in case of Rydet). Please note that although some of their depicted "protagonists" in the spectacle of a society may seem weird, none of them is weird in a more disturbing, psychologically "wicked" way in which Diane Arbus photographed her "heroes" and "heroines" in the USA of the 1960s.

In order to appreciate these "layers" of meaning in a photograph, which should not be taken at the face value of a documentary report, one should look toward a more qualitative methodology in research preferences. Sztompka can be forgiven for his simplistic approach toward the images skillfully constructed by Rydet, because he is not the only one who had missed the beat of:

> the exploding culture in the sway of painters, dealers, critics, shopkeepers, second sons, Russian epicures, Spanish parvenus, and American expatriates. Jews abounded, as did homosexuals, bisexuals, Bolsheviks and women in sensible shoes. (…) And most disturbingly, for those who felt they ought to be in control – or that someone should be – beauties proliferated, each finding an audience, each bearing its own little rhetorical load of psychopolitical permission. (Hickey 2009, p. 61)

Exchange "beauties" in artistic criticism for "paradigmatic schools" in research communities and you will see why some sociologists, especially if they are tainted by the vain hopes of ever approaching the neo-positivist ideal of strict scientific methodology, cannot see beyond a documentary—the symbolic message is lost on them. The temptation is very strong, because one of the first populist cultural campaigns, which used photography as a weapon of mass communication was Andre Malraux's "imaginary museum" of world visual art works, supposedly accessible to all mankind through increasingly accurate photographic reproductions. Malraux, whose ideals of the social functions of art had been shaped by the leftist policies of the cultural commissars of the "united working class popular fronts" of the 1930s (the Spanish Civil War was the rallying cry), saw the role of photography allowing to disseminate copies of works of art to every nook and cranny of industrial society as a huge opening of the gates to the "gardens of culture", heretofore reserved only for the rich and the educated. Imaginary museum was a virtual collection which never closed to the mass audience. Dreams of a liberating role of new technologies die hard, in spite of the fact that a quarter of a century before "imaginary museum" had been conceived by French minister of culture, an elitist critic of the role of a work of art in the era of its mechanical reproduction regretted the loss of "aura" in works of art disseminated as printed photographic reproductions of originals (Walter Benjamin). But let us turn to more qualitatively minded research communities.

4.4   Photography in Moderate Qualitative Paradigmatic Climates

Krzysztof Konecki, editor in chief of the *Qualitative Sociology Review*, also does, like his older colleague, Piotr Sztompka, make use of photographs in his research projects—but his methodological constraints on their use indicate a much more sophisticated level of visual analysis. For instance, when analyzing the photographs made of household pets, most commonly cats and dogs, he points out that the ongoing "anthropomorphization" of household pets (who become symbolic "members of the family") is built by many gestures, body postures, non-verbal and paraverbal communication, all three of which are particularly well traceable in the construction of "social spaces" by owners of pets, who make their photographs as intended symbolic products "of", not only "about" a culture.

Konecki is interested in a multi-layered analysis of "slices" of meanings encoded in photographs as documents "of" cultures (he introduces another term, apart from visual anthropology and sociology, namely "visual ethnography"). He points out that we are increasingly required to process numerous visual artifacts as we pursue our daily routines, as we translate our verbal communications into PowerPoint presentations with multiple visual aids, when we record with video and photographic cameras, when we send photographs through our mobile phones, indulge in running or visiting blogs, and so on. Recommending an alliance of "Grounded Theory Plus Visual Data" (the title of his unpublished paper), the sociologist of working organizations and urban life writes:

> If grounded theorists want to create substantive theories about contemporary world, they have no choice, they have to turn to visual data and visual processes, which are a vital component of actions within contemporary institutions and our social world in general. Visual data open new possibilities to develop grounded theories. Developing of theories of substantive visual processes could facilitate constructing formal theories of visualization of social problem, visualization of organizational politics, visualization of identity, etc. The most ambitious goal looms large on a theoretical horizon: the construction of a formal theory of visualization of action. Future of grounded theories will be inevitably associated with generating theory on social, cultural, and psychological dimensions of visual reality, not only because of our society's recent "visual turn" but also because of growing research focus on visuality of our social worlds. (…) It is time to turn to "unspeakable" phenomena, which cannot be directly communicated – verbally. (Konecki 2005, p. 21)

The same qualitative sociologist in search of a grounded theory built with visual data thus promotes a research methodology based on photographic records, whose execution is subjected to a rigorous methodology, which allows a camera-wielding sociologist to understand the role of images, especially photographic images, in the broader process of the "theatralization of social space" (Magala 1978, 2009). Thus Konecki would have incorporated "Sanderian" or "Rydetian" projects into his research methodology as sophisticated symbols "of" their respective cultures and contributions to the ongoing negotiation of continually constructed and reconstructed social realities. However, communicative games do not wait for visual sociologists or visual anthropologists, or even visual ethnographers to hone their methodological tools and to tackle them—they evolve and become more sophisticated, more resistant to the critical questioning within the methodological research programs of yesterday. They are also being solved—or at least publicly discussed outside of the realm of science and scholarship. What should be done to stay au courant with the evolving societies where all that is solid melts into thin air (Halvani 2010)?

One way of staying ahead is to go for a broad alliance of sociologists (who focus on the menu of interactions sprinkled with institutional sauce), cultural anthropologists (who will furnish the horse-d'oeuvres of individual actions in psychosocial salads), cultural scholars (who will season the main dish with culturally legitimized valuation) and media experts (who will see to it that the meal comes out as it was intended to or at least does not taste entirely differently than intended, expected, experienced). This is the solution suggested by the visual sociologists, who want to study photography and her images both on and behind the social stages, on which they are displayed: They want to study

> [a] broad spectrum of different micro-practices connected, primarily, to the emergence of photographic images, and second, to the methods of managing photographs (one's own, but also those, which had been inherited, donated, acquired, etc.), third, to the ways of incorporating them in (or excluding them from) private stage designs constructed by individuals in order to live and experience daily life. (Drozdowski and Krajewski 2010, p. 4)

Photography (but also other forms of multimediated motion pictures communications—video, visual blogs, facebook photographic and video records, etc.) thrives in these more qualitatively oriented paradigmatic climates of (visual) anthropological and sociological research communities—but the relationship between photography as the preferred supplier of subcomponents to the artists (whose responsibility is the final assembly line of the work of art) and photography as the privileged supplier of sophisticated visual symbols of a culture to communities of choice (membership in which can become a matter of hot debate, cf.Kanter 1977) can become blurred at times. When does art precede science and scholarship, demonstrating possible future research strategies?

4.5 Visual Interfaces Between Art, Research and Management: Cases in Point

4.5.1 Open/Closed Performance/Installation/Visual Research Project by Jan Piekarczyk (Summer 2006, Warsaw)[1]

In the summer of 2006, between June 28 and August 22, the resident of Warsaw and a photographer/performer, Jan Piekarczyk, went out and asked 1000 inhabitants of Warsaw who happened to be passersby, if he could take a picture of their faces. If a person agreed, he would make a color photograph, with clearly visible facial features. If a person refused, he would not record the face, but take a black-and-white picture of an anonymous detail—a piece of an elbow, a tip of a shoe, or a shadow on a pavement. Having assembled 1000 of photographs in this way—he had opened an exhibition "Open/Closed", which had been composed of a huge tableau made of 1000 images, some of them with colored portraits and some with black-and-white details (if sometimes he did—by coincidence—register a face of the refusing individual, he would mask it with black strip in order to render the face anonymous). What were the proportions? In this particular case 488 individuals agreed to have their portrait made and their face exhibited in the subsequent exhibition. This means that 512 refused to agree, sometimes with a display of aggressive behavior. The visual artist, member of the board of Society for the Artists of Other Arts and the editor and publisher of an online quarterly "Galois" gave it a modest title "Performance for 1000 Individuals, Camera and Myself". Sometimes people approached by Piekarczyk had reflected on their automatic behavior and changed it. The artist described, for instance, the incident with two young, attractive women, one Turkish, one Polish, both waiting on a bus stop at the center of Warsaw. Both were close enough for the Turkish one to hear that the Polish one did not agree to have her picture taken, and for the Polish one to hear a few seconds later that the Turkish one agreed at once. Piekarczyk had photographed the Turkish one and then returned to the Polish woman asking her again. The young woman started thinking aloud and asked herself why, indeed, did she give a negative response the first time round. Having found no good reasons, she changed her mind and allowed the artist to photograph her face. Perhaps, if we are to believe the artist, she had felt that for the Turkish woman the act of agreement was a much braver decision—to open her face to the camera meant not only allowing a gaze of the others run freely and examine its details. This open exposure to naked gaze also ran against the grain of the Muslim veil, it may have had some rebellious undertones of a liberating, defying escape from the obligatory closure and from hiding beneath a parda, a chador, a kvef or a veil.

The artist kept a meticulous record of his actions: he approached all individuals, explaining that he was executing "an independent artistic project", avoided those in a hurry, those sunk in conversation, those who suspected him of selling or advertising something or "marginalized" figures (beggars, homeless, pickpockets). The proportions of refusals and agreements (48.8%—"open" ones, 51.2%—"closed" ones) prompted the artist-performer-researcher to comment on "openness" and "closure" and their significance for a "photography of a psychological climate of a city". He kept some statistical data per each 100 respondents:

1. The first 100, 47%—open, 53%—closed
2. The second 100, 52%—open, 48%—closed
3. The third 100, 51%—open, 49%—closed
4. The fourth 100, 39%—open, 61%—closed
5. The fifth 100, 51%—open, 49%—closed
6. The sixth 100, 50%—open, 50%—closed
7. The seventh 100, 45%—open, 55%—closed
8. The eighth 100, 38%—open, 62%—closed
9. The ninth 100, 55%—open, 45%—closed
10. The tenth 100, 60%—open, 40%—closed

Let us examine his own comments on the results of his project:

> My project is an artistic one. It is a conceptual-sociological genre of art, with some elements of performance—when I am approaching individuals, asking them questions, and actually making a photograph. Statistical data are a byproduct. The "open" ones are those who agree to be photographed. They accept themselves. The "closed" ones, are those who do not agree. The "closed" ones are closed for various reasons, in contingent circumstances (living contexts). They have no self-ironic distance to themselves and some of them probably do not accept, do not approve of themselves. (Piekarczyk, in a radio interview, September 2006)

It would be tempting to consider a similar action-performance-project carried out simultaneously in various urban centers all around the globe—but this is easier said and done in academic communities (which are already globally networked on the research level and regionally networked on the teaching level) than in the artistic ones (which are only imperfectly linked by the commercial networks of major art fairs and major exhibiting events in cultural politics—Documenta in Kassel, Biennale in Venice or major traveling exhibitions are cases in point). Perhaps we shall have some day a real global grass-roots alliance and networking of freely clustering artists and their indispensable fellow networkers (curators, critics, theoreticians, empirical researchers, politicians, experts in cultural policies, aestheticians, media professionals, journalists, columnists, spin doctors, educators, young aspiring students of arts and sciences, advertising gurus, wealthy and average aspiring collectors, TV performers, cultural entrepreneurs, etc.). At present, the Open/Closed project of Jan Piekarczyk from 2006 remains a relatively local and unexplored example of a successful, robust, fertile, promising, creative, imaginative—both aesthetically and cognitively—performative project executed at the interface of visual anthropology, sociology or ethnography on the one hand and visual arts on the other.

Perhaps it could—one day—become a pioneering influence upon systematic encouragement of new "communicative powers" (to paraphrase Castells), of new communicative and interactive "social technologies" in service of constructing new public spaces for enhanced civic communications, for (re)negotiating our "sociabilities" (abilities to socialize) and for replenishing our social "imaginaries" (Taylor 2004).

4.5.2 Visual Narratives Between Touchscreens and Communities

Visual anthropology and visual sociology are challenged by the new technologies of visual communications. Blogs have to be studied (Olechnicki 2009), facebook accounts have to be analyzed, online dating has to be explored, and personal connectivity has to be investigated. Visual anthropologists and visual sociologists respond with attempts to go beyond the classical sociology or beyond the qualitative methodologies relying mostly on verbal interpretations and tend to cluster around the so-called narrative methodologies, which increasingly often include the visual elements.

David Boje's explorations of the narrative strategies of Nike, McDonald's and Disney are cases in point (Boje 2008). It is no coincidence that Boje develops his theory of sensemaking in organizations with the aid of "living stories", "dead narratives" and "ante-narratives" quoting among others—Walter Benjamin and Gertrude Stein, both of whom were keenly aware of the increasing role of visual "clues" in our communications. Visual sociologists and anthropologists (and organizational ethnographers in schools of business) are also challenged by a changing way in which we conceptualize images as they enter our communications, influence our interactions and shape the visions of virtual and real futures. The question posed with brutal directness and simplicity by Mitchell—"What do pictures want?"—is thus legitimate, justified and relevant. We guarantee its salience, because we do want to "know". We want to "see" (the expression "I see" is equivalent to "I understand" in most languages). We want to know what pictures want from us. We want to know what we want from the others when "speaking" through images to them. Finally, we want to know, more generally, what branding of cultural contents in creative industries means for ethics and aesthetics (hence the emergence of "creative industries observatories" in London, Los Angeles or New York). Paraphrasing Freud, who asked the same question, allegedly, about women ("what do women want?"), we investigate images as our extensions, our protheses, but also illusions, testimonies to lost dreams, or, as McLuhan once phrased it "amputations". We want to know how ethics (measuring our actions against the expectations and preferences of captive audiences) and aesthetics (evaluating experiences triggered by images we disseminate) are related to one another. We want to know how they are **intertwined** in the structures and processes of communications. Since the domain of social communications is growing to immense proportions (we are hyper-connected and mobile and suppliers of pure connectivity are earning many times more than suppliers of cultural contents), we have to focus on some samples, some selected sections, which we can still relatively easily isolate and study. This is the basic explanation for the career of the "**artworld**" as an area of studies performed by visual anthropologists, visual sociologists, visual ethnographers and other academic specialists. Visual sociologists are thus assisted by art critics, art historians, art disseminators, cultural experts (curators, gallery owners, museum directors). Why artworlds? Because they are smaller than all socially organized communications and because they offer a chance of studying aesthetic values emergent from individual and collective interactions. Hence an emergent stress on interactivity in performative arts. Hence the focus on both stage arts and performance artists recruited from among poets, sculptors, painters or installation authors, say both Baryshnikov or Gielgud on the one hand and Marina Abramovic or Christo on the other. Hence the focus on "dialogue", on "feedbacking" in institutional encounters with audiences and constituencies and the entire new wave of "relational aesthetics" (cf. Bourriaud 2009), which came about as a result of a conscious resistance to social formatting and standardization.

A quick click brings us to the Wikipedian definition of an artworld as a loosely related network and overlapping organizations of individuals who produce, commission, preserve acknowledged works of art and also sell and buy "works of art". Two definitions are usually quoted, first the classical one by Howard Becker first coined in 1982:

> the network of people whose cooperative activity, organized via their joint
> knowledge of conventional means of doing things, produce(s) the kind of

art works that art world is noted for This tautological definition mirrors the analysis, which is less a logically organized sociological theory of art than an exploration of the potential of the idea of an art world for increasing our understanding of how people produce and consume art works.(...) the principle of analysis is social organizational, not aesthetic. (Becker 1982, pp x–xi)

Second, a definition by a contemporary empirical analyst, a practicing female critic and a participating observer touring all major art sales, exhibitions and auctions, namely, Sarah Thornton. According to her, an art-world is "a loose network of overlapping subcultures held together by a belief in art. They span the globe but cluster in art capitals like New York, London, Los Angeles, and Berlin" (Thornton 2008).

Let us note the presence of an act of faith in Thornton's recent definition of an artworld—as opposed to the attempted purging of an aesthetic reflection from a social organizational analysis in Becker's approach. Social organization but also a leap of faith are being accounted for. A leap of faith (or an assumption that studied individuals have made a great spiritual leap forward toward a transcendent value) is required in order to "make sense" of an artworld. In this the artworld resembles the world of religion, the "churchworld". Very much like churches would not have made sense if there were no believers inside them, if individuals did not have faith in God, musea and art galleries would not have made much sense either, if there were no individuals convinced that works of art do allow for valuable experiences, which transcend the material calculus of utilitarian pleasures and open up new, more meaningful experiences of being human, being together with the others, deciding about our common futures. Hence nobody misses a beat when sociological and aesthetic studies are published under such titles as "Theology Goes to the Movies" (Marsh 2007) or "The Enchantment of Modern Life" (cf. Bennett 2001). The problem is with the level of sophistication of this faith, which has to survive Duchamp, Warhol, David Hockney's iPhone "paintings" and an ongoing hum of unrelenting criticism of everything by everybody:

I believe that magical attitudes towards images are just as powerful in the modern world as they were in the so-called ages of faith. (...) My argument is that the double consciousness about images is a deep and abiding feature of human responses to representation. It is not something we "get over" as we grow up, become modern or acquire critical consciousness. (…) The specific expressions of this paradoxical double consciousness of images are amazingly various. (…) They include the ineluctable tendency of criticism itself to pose as an iconoclastic practice, a labor of demystification and pedagogical exposure of false images. Critique-as-iconoclasm is, in my view, just as much a symptom of the life of images as its obverse, the naive faith in the inner life of works of art. (…) We might have to entertain what I would call a "critical idolatry" or "secular divination" as an antidote to that reflexive iconoclasm that governs intellectual discourse today. Critical idolatry involves an approach to images that does not dream of destroying them and that recognizes every act of disfiguration or defacement as itself an act of creative destruction for which we must take responsibility. (Mitchell 2005, pp. 8, 26)

The continuing existence of both churches and musea testifies to the attractiveness of the (visual or visually enhanced) "narratives" told by religious or artistic story-tellers. Visually enhanced narratives prosper in spite of two serious blows to the ontology of the works of art (analogy to the ontological arguments about God's existence and articles of faith is striking). I am referring to the blow executed by Duchamp (any object can function as a work of art replacing it in a unique context of an avant-garde exhibition), Warhol (any mass-produced reproduction of non-exceptional objects can function as a work of art in "imaginary museum") and Hockney (any virtual composition can function as a work of art between digitally communicating individuals). The sequence of important events in defining contemporary artistic practices—and the one that visual anthropology, sociology and ethnography should study following developments in visual arts and visual communications—is as follows:

- In 1917, Marcel Duchamp, using the assumed name of "R.Mutt", submitted a sculpture to the Armory Show in New York City. The sculpture consisted of a standard porcelain "urinoir" purchased in the construction supply warehouse and a socket, on which it had been mounted. The title of this work of art submitted to the exhibition was "The Fountain" (as a matter of fact the organizers did not dare to expose it, but Alfred Stieglitz had already photographed it for the catalogue and so Duchamp's gesture had been publicly acknowledged entering the public negotiation of the "borders" of art). He succeeded in persuading the artworld that an intention of the artist is more important than any material embodiment of his or her idea (and that the "core" of artistic communication cannot reside in purely material properties of "unique objects").

– In 1962 Andy Warhol started to produce paintings by transferring silk-screen images onto the canvas. He was using the same technology of cheap printing, which had been used to mass-produce T-shirts and other "souvenir objects". His "Factory" continued to work until the 1980s. Production was limited only by what Warhol thought he could sell at the time, not by production capacities of his assistants from the factory – who mostly executed "art by phone", namely following Warhol's instructions, although the artist might not physically have been present at the moment the "painting" on canvas had been made.

– In 2009, the painter David Hockney started producing images on the touchscreen of his iPhone moving his thumb and activating the "paintbrush" function. Upon completion, he would click the automatic e-mail and the image would be sent to more than a hundred of his friends and acquaintances. We are dealing with a direct transmission of the artist's intention to his viewers, "consumers" of the image he had just produced. This communication happens without any interfering "original", which has to be processed and presented by professional experts. No intermediaries are needed, since the original object of art is a virtual constellation of digitally operated liquid crystals "recorded" only in the virtual spaces and memories of electronic devices. Of course, one still needs historians and teachers to make individuals aware of the value of what Hockney is sending by e-mail, but they would have done their work when the above communication actually takes place.

This is the story of conceptual and technological shortening of a distance between an artist and individuals willing to regard what he or she has done and willing to share the designed experience. But it is not the entire story. For a visual anthropologist and sociologist, the most interesting aspects of these developments are happening among the "distributors" of aesthetic communications (for a diamond model of artistic communications, see Alexander 2003), or "go-betweens" connecting arts and communities, creators and "consumers". The controversial decision of Andy Warhol Foundation to "deny" authenticity of one of the best-known Warhol "prints on canvas" is the case in point. In 2003, one of the best known Warhol "red" self-portraits, silk-printed on canvas, framed and signed personally by the artist, who gave it to his friend, Bruno Bischofberger, in Zurich in 1969, was stamped on the backside frame by the Warhol Foundation experts with the read letters "DENIED". How could the foundation deny authenticity of a work, which had been personally signed by the artist, personally dedicated to a friend and personally given to him with many witnesses around?

Moreover, Warhol had deliberately chosen it for the cover of the "catalogue raisonne" of the only major retrospective exhibition that he had organized himself (in ICA of Philadelphia, in October 1965). The argument of the foundation experts was that Warhol was only giving instructions on the phone to his collaborators, but he had not been physically present in "the factory" when the printing of canvas was actually "finished". They refused to acknowledge the explanations of Warhol himself, who argued that he would like to move away from a personal touch in painting. Their experts issue a certificate of authenticity only to the owners of those prints, which had been produced earlier, when Warhol did not yet master the entire process so well as to be able to limit himself to the telephone instructions. Is this a valid argument? There are serious doubts about the soundness of this decision. First, the intention of the artist to author this particular work was clearly present, second, control over the production process did not require physical presence at the factory for the finishing phases of production, third, ever since Renaissance the painter's signature and even more so a dedication to a given individual scribbled on the back of the painting were enough to "authorize" the work of art. But there is more: the Warhol Foundation owes about $500 million worth of artworks by Andy Warhol and by controlling the market (for instance by limiting the value of works owned by other parties through denial of an official "authentification") stands to make a lot more money than it would otherwise have done. The foundation is already challenged in a growing number of court cases, in which other collectors charge it with illegal restricting of the number of art works by Warhol on the market in order to increase the value of their own holdings. "Decisions, like the one about 'Bruno B. Self Portrait' at best raise doubts about this board's competence and at worst about its integrity."(Dorment 2009)

4.5.3 "Regarding the Pain of Others" or Visual Anthropology of Political Marketing

The "Warhol wars" described above are mostly about marketing of a very large ("industrial scale") body of work, which has been left after the artist's death and is still being "fed" to the disseminating organizations or gallery/auction/direct approach selling points. Serious conflicts and clashes of different views on what exactly was a legitimate creative activity and how completely can a work of art "come unmoored" (cf. Irvin 2009) from physical presence of the "authorizing artist" still leave us outside of the dangerous zone where body count of dead and wounded is performed as a matter of routine. This is not the case with the visual impact of "regarding the pain of others", as Susan Sontag had phrased it in the last collection of her essays on functions of visual reports on atrocities in our societies.

In order to analyze these functions, let us compare three different but closely related visual communications: the documentary photograph (by the Vietnamese Hyunh Cong Ut), the photographic work of art (by the Polish Zbigniew Libera), and the political work of visual arts inspired by a documentary photograph (by the British Bangsy).

The first had been made in 1972 and the girl registered in the center of the photograph was actually suffering from napalm burns. The photographer took the picture, then put his camera aside, picked up the girl and ran with her to the medics to have her treated for her wounds. He saved her life and Kim Phuc is presently living in Canada, where she is, among others, a peace activist. This differs from behavior of another photographer, Tyler Hicks, who had witnessed a lynching of a Taliban prisoner of war in the hands of the Northern alliance's soldiers in Afghanistan in

2001. His documentary photographs make it clear to the viewers that he had recorded successive phases of the killing without any attempt at either saving the prisoner's life or reducing his suffering. Likewise, after the Haitian earthquake of 2009, the cameraman of Fox television in Port-au-Prince filmed a Haitian who had been shot by the police when plundering a supermarket and lay dying on the street. The cameraman continued filming until the shot Haitian bled to death, making no attempt either to call medical assistance or to try to stop bleeding himself. In view of these two later cases, one could be skeptical about the producers of photographic imagery—somehow the document recorded by Hyunh Cong Ut seems to us more "convincing", more "honest" than the other two reports, which had been made at a high price of suppressing one's compassion and solidarity with the suffering others. But the role of the original document from Vietnam War had already been played; we know that it had contributed, together with many other documentary reports on the suffering and human cost of the military actions, to the ultimate negotiations and the end of the Vietnam War.

Long after the end of the war, however, the photograph in question became a "secular icon", an element of a collective memory of the horrors of war. Zbigniew Libera, the Polish visual artist (better known from his Lego box for constructing a do-it-yourself concentration camp), had picked this iconic image as a collective "negative" memory, a "meme" of our visual heritage, as a symbolic "relic" in our photographic tradition. He had decided to turn it upside down and to rebuild it in collective memory—but this time with an optimistic happy end, with a "plus" instead of a "minus", as a "positive" memory, not a negative one. Hence a dramatic tragedy of a group of children escaping a napalm bombing of their village turns into an idyllic outing of a group of friends who had been exercising parachute springing and enjoying the open space, wide horizons, green fields and smiling companions. As they return, the girl, who is still a central figure in the composition, is still naked, is still running and is still in the middle of the country road—but this time she is also smiling, her nakedness suggests a swimming episode a while ago rather than a horror of burning clothes and a general being at ease with her body, as if she were a bucolic nymph. And more generally—Libera appears to be asking—do we really know that things happened the way the icon in our collective memories represented it? How can we be sure that the records of real life event were not tampered with? How can we be sure that nobody re-arranged the records in order to get some message through to us?

Libera's intentions are not political: he does not want to criticize any of the parties fighting the Vietnam War, nor does he want to reconsider the responsibilities of the combatants for suffering inflicted upon civilians. Bangsy, who places anonymous satirical, critical murals on the houses and walls in the streets of London, clearly does have a political point to make. He has also picked up the Vietnam photograph by Hyunh Cong Ut, but he had turned it into a comic strip populated by the iconic "logos" of the US multinationals—Mickey Mouse stands for Disney corporation and Ronald McDonald stands for the McDonald's corporation, and both stand for the US capitalist economy, which dominates the world markets. The superheroes are supposed to be children-friendly and this is how they are depicted in the media—so that they hold hands of our Vietnamese girl in a gesture of joyful innocence and "happy hour" for the entire world. At the same time the girl from the original photograph is stark naked—which in itself is against the principles of either Disneyland's or McDonald's restaurants—and apparently crying (no wonder, in the original documentary record she had just been sprinkled with burning substance). So Bangsy's message is loud and clear—after having been injured by the real US troops in Vietnam, the girl is symbolically "saved" and offered consolation by fictitious American heroes, by mythical, non-existing figments of popular imagination and mass consumption. Damage is real, consolation prize—illusory. Globalization à la "the West and the rest" is not good for little girls (nor—by ideological extension—for anybody else). This is a politically committed message, an ideological manifesto. But does it really prompt an action? If so, what action could it be? If not, does it not simply add more cynicism to the eyes of the passive consumers of fresh images? Doesn't Bangsy, through his cynicism and lack of clear-cut action plan, contribute to the domination of "mental capitalism" he attempts to satirize?

The war in Vietnam is not considered a purely imperialist, one-sided aggression of the US troops against the peace-loving south Vietnamese peasants anymore. It is considered to have been one of the major wars by proxies conducted in the period of the Cold War. The northern Vietnamese communists had won because they did not hesitate to sacrifice more civilians (both in the north and in the south) and military than the USA (and their ally, the south Vietnamese government). They had also won, because there was no such thing as the backlash of public opinion—the Hanoi communists were not afraid of becoming less popular before the next elections. The consolation offered by popular culture icons may be illusory (and the multinational corporations are often subjected to public criticism in the media of western democracies), but so was the one offered by Lenin and Ho Chi Minh (and in spite of the shift toward the market economy, the Vietnamese society has yet to see the evaluation of the responsibility of communist elites for the bloody foreign war waged beyond the southern border). Visual communications do play a role in this process, as Sontag was crucially aware of (although she had made a politically correct leftist trip to Hanoi and, later on, to Sarajevo):

> The familiarity of certain photographs builds our sense of the present and immediate past. Photographs lay down routes of reference, and serve as totems of causes: sentiment is more likely to crystallize around a photograph than around a verbal slogan. And photographs help to construct – and revise – our sense of a more distant past, with the posthumous shocks engineered by the circulation of hitherto unknown photographs. Photographs that everyone recognizes are now a constituent part of what a society chooses to think about, or declares that it has chosen to think about. It calls these ideas "memories" and that is, over the long run, a fiction. Strictly speaking, there is no such thing as collective memory – part of the same family of spurious notions as collective guilt. But there is collective instruction. (Sontag 2003, p. 85)

The term "collective instruction" might suggest some structured and synchronized effort—similar to the efforts of "hidden persuaders" in the world of commercial advertising and marketing, or even "visible persuaders" as was the case with the propaganda empires of Stalin or Hitler. However, we are all subjected to a collective instruction from a variety of sources and agencies. At the same time, there is no centrally acknowledged authority, which would check the quality of our individual "mixes" of media influences, or which would grade our "homeworks" ("The way you had organized your memories of grandparents banned to Kazakhstan is great and could be an example for the others") and publish regular "rankings" of our collective memories (e.g. Katyń up in the EU, Soveto down in the USA). Collective instruction is something public intellectuals diagnose—not in the least with the research methods of visual anthropology, ethnography and sociology.

Notes

1. I thank Ms Ewa Mikina for pointing out this particular artistic project to me.

References

Alexander, V. D. (2003). *Sociology of the Arts. Exploring Fine and Popular Forms.* Malden/Oxford: Blackwell.

Ankere, S., & Nelkin, D. (2003). *The Molecular Gaze. Art in the Genetic Age.* Cold Spring Harbor: Cold Spring Harbor Laboratory Press.

Archer, M. (2007). *Making Our Way Through the World. Human Reflexivity and Social Mobility.* Cambridge/New York: Cambridge University Press.

Barthes, R. (1980). *La chambre claire.* Paris: Seuil.

Becker, H. S. (1982). *Art Worlds.* Berkeley/Los Angeles/London: University of California Press.

Belting, H. (2003). *Art History After Modernism.* Chicago/London: The University of Chicago Press.

Bennett, J. (2001). *The Enchantment of Modern Life. Attachments, Crossings, and Ethics.* Princeton/Oxford: Princeton University Press.

Boje, D. (2008). *Storytelling Organizations.* Los Angeles/London/New Delhi/Singapore: Sage.

Bourriaud, N. (2009). *The Radicant.* Berlin/New York: Sternberg Press.

Butler, J. (2006). *Gender Trouble: Feminism and the Subversion of Identity.* London/New York: Routledge. (The First edition – 1990).

Castells, M. (1996). *The Network Society (First Part of the Trilogy: The Information Age: Economy, Society and Culture).* Oxford/Malden: Blackwell.

Castells, M. (2001). The Internet Galaxy. In *Reflections on the Internet, Business, and Society*. Oxford/New York: Oxford University Press.

Castells, M. (2009). *Communication Power*. Oxford/New York: Oxford University Press.

Costello, D., & Willsdon, D. (Eds.). (2008). *The Life and Death of Images. Ethics & Aesthetics*. Ithaca: Cornell University Press.

Crimp, D. (1997). *On the Museum's Ruins*. Cambridge, MA/London: The MIT Press.

Dikovitskaya, M. (2005). *Visual Culture. The Study of the Visual after the Cultural Turn*. Cambridge, MA/London: The MIT Press.

Dorment, R. (2009, October 22). What Is an Andy Warhol? *New York Review of Books, 56*(16), pp. 23–25.

Drozdowski, R., & Krajewski, M. (2010). *For Photography* (Za fotografię, in Polish). Warsaw: Korporacja Ha!art.

Duve, T. (1998). *Kant After Duchamp*. Cambridge, MA/London: The MIT Press.

Elkins, J. (Ed.). (2007). *Photography Theory*. New York/London: Routledge.

Ewen, S. (1988). *All Consuming Images. The Politics of Style in Contemporary Culture*. New York: Basic Books.

Foster, H. (2004). *Prosthetic Gods*. Cambridge, MA/London: The MIT Press.

Franck, G. (1998). *Oekonomie der Aufmerksamkeit*. Vienna: Hanser.

Franck, G. (2005). *Mentaler Kapitalismus. Eine politische Oekonomie des Geistes*. Vienna: Hanser.

Goffman, E. (1988). *Gender Advertisements*. New York: Harper Collins. (The First edition – 1976).

Halvani, R. (2010). *Philosophy of Love, Sex and Marriage. An Introduction*. New York/London: Routledge.

Hariman, R. H., & Lucaites, J.-L. (2007). *No Caption Needed. Iconic Photographs, Public Culture and Liberal Democracy*. Chicago: The University of Chicago Press.

Hickey, D. (2009). *The Invisible Dragon. Essays on Beauty*. Chicago/London: The Chicago University Press.

Hindletter, B., Kaizen, W., Maimon, V., Mansoor, J., & McCormick, S. (Eds.). (2009). *Communities of Sense. Rethinking Aesthetics and Politics*. Durham/London: Duke University Press.

Irvin, S. (2009). The Ontological Diversity of Visual Artworks. In K. Stock & K. Thomson-Jones (Eds.), *New Waves in Aesthetics*. Basingstoke/New York: Palgrave Macmillan.

Kanter, R. M. (1993). *Men and Women of the Corporation*. New York: Basic Books. (The First edition – 1977).

Konecki, K. (2005). *Ludzie i ich zwierzęta. Interakcjoinistyczno-symboliczna analiza świata społecznego zwierząt domowych* [People and Their Animals. Interactionist-Symbolic Analysis of Social Life of Domestic Pets]. Warszawa: Wydawnictwo SCHOLAR (in Polish).

Levin, D. M. (Ed.). (1993). *Modernity and the Hegemony of Vision*. Berkeley/Los Angeles/London: University of California Press.

Magala, S. (1978). Photography as an Element of the Theatralization of Social Life (Fotografia jako element teatralizacji życia społecznego). *Fotografia 4*, 6–7 (in Polish).

Magala, S. (2009). *The Management of Meaning in Organizations*. Palgrave Macmillan: Basingstoke/New York.

Manovich, L. (2001). *The Language of the New Media*. Cambridge, MA/London: The MIT Press.

Marsh, C. (2007). *Theology Goes to the Movies. An Introduction to Critical Christian Thinking*. London/New York: Routledge.

McLuhan, M. (1964). *Understanding Media*. New York: New American Library.

Mitchell, W. J. T. (2005). *What Do Pictures Want? The Lives and Loves of Images*. Chicago/London: The University of Chicago Press.

Olechnicki, K. (Ed.). (2003a). *Obrazy w działaniu. Studia z socjologii i antropologii obrazu* [Images in Action. Studies in Sociology and Anthropology of Images]. Toruń: Nicolaus Copernicus University Publishing House. (in Polish, with English abstracts).

Olechnicki, K. (2003b). *Antropologia obrazu. Fotografia jako metoda, przedmiot i medium nauk społecznych* [Anthropology of Images. Photography as a Method, Object and Medium of Social Sciences]. Warsaw: Oficyna Naukowa (in Polish).

Olechnicki, K. (2009). *Fotoblogi.Pamiętniki z opcją przekazu. Fotografia i fotoblogerzy w kulturze konsumpcyjnej* [Photoblogs. Diaries with Communicating Option. Photography and Photobloggers in the Culture of Consumption]. Warsaw: Wydawnictwa Akademickie i Profesjonalne (in Polish).

Panofsky, E. (1982). *Meaning in the Visual Arts*. Chicago/London: The University of Chicago Press. (The First edition – 1972).

Pauwels, L. (2006). *Visual Cultures of Science: Visual Representations and Expression in Scientific Knowledge Building and Science Communication*. Hannover: Dartmouth College Press.

Pauwels, L. (2009). Visual Literacy, Visual Culture and Visual Scholarship: Adjusting a Distorted Picture. In *Proceedings of IVLA*, pp. 19–24.

Ranciere, J. (2009). *The Emancipated Spectator*. London/New York: Verso. (The First French edition – 2008).

Sikora, S. (2004). *Fotografia między dokumentem a symbolem* [Photography Between a Document and a Symbol]. Warszawa/Izabelin: Świat Literacki/ Instytut Sztuki PAN (in Polish).

Sontag, S. (1977). *On Photography*. New York: Farrar, Straus & Giroux.

Sontag, S. (2001). *Where the Stress Falls*. New York: Farrar, Straus & Giroux.

Sontag, S. (2003). *Regarding the Pain of Others*. New York: Farrar, Straus & Giroux.

Sztompka, P. (2005). *Socjologia wizualna. Fotografia jako metoda badawcza* [Visual Sociology. Photography as a Research Method]. Warszawa: Wydawnictwo Naukowe PWN (in Polish).

Taylor, C. (2004). *Modern Social Imaginaries*. Durham/London: Duke University Press.

Thornton, S. (2008). *Seven Days in the Art World*. London: Granta.

Turkle, S. (1995). *Life on the Screen. Identity in the Age of the Internet*. New York: Simon & Schuster.

Wasik, B. (2009). *And Then There Is This. How Stories Live and Die in Viral Culture*. New York: Viking.

Wells, L. (Ed.). (1997). *Photography. A Critical Introduction*. New York/London: Routledge.

Worth, S. (1981). *Studying Visual Communication*. (Ed. Larry Gross). Philadelphia: The University of Pennsylvania Press

5

Action Research

Davydd J. Greenwood

5.1 Introduction

Action research (AR) is a strategy for social research that combines the
expertise and facilitation of a professional social researcher with the
knowledge, energy, and commitments of local stakeholders in a particular organizational, community, political, or environmental setting.
Together, these actors form a collaborative learning community to define
the problems, decide the data needed to understand them, and generate
hypotheses about the relevant causes. They then engage together in gathering data, recruiting additional stakeholders, and interpreting the results.
Finally, they co-design the actions arising from their results to ameliorate
the problems, take the actions, and then evaluate the results. They evaluate the results together, and if the results do not meet their expectations,
they engage in further cycles of research, analysis, and action until the

D.J. Greenwood (✉)
Cornell University, Ithaca, NY, USA

© The Author(s) 2018
M. Ciesielska, D. Jemielniak (eds.), *Qualitative Methodologies in Organization Studies*,
https://doi.org/10.1007/978-3-319-65217-7_5

problems have been addressed to their satisfaction. The learning community so created operates according to a set of values that privilege respect for the knowledge and interests of all participants (including the social researcher), democratic dialogue that aims to permit the group to learn from the experiences and commitments of all of its members, and that is premised on the ability of all people to become more effective researchers and to act more successfully on their own behalf. Action research is guided by value commitments that include enhancing democratic participation, increasing people's ability to pursue their own interests, educating non-professional researchers in the use and critique of techniques of social research and in the wise use of professional consultants.[1]

5.2   What Action Research Is Not

Conventional social researchers divide themselves generally into "quantitative" researchers who use numbers and statistical models and "qualitative" researchers who privilege interpretive and symbolic approaches to research. Action researchers reject making such a choice on pragmatic grounds. Our choice of methods and approaches must be dictated by the requirements of the problem being addressed. If an oil spill has polluted a domestic water supply and the oil company denies it, then a quantitative scientific analysis of the geology, groundwater, and related matters is a central part of the research. If public officials are not enforcing zoning laws, satellite imagery, GIS data, tax assessment rolls, and so on form part of the research response. If non-native speakers of the official language in public schools are being patronized and not taken seriously as students, in addition to the educational outcomes data, ethnographic analyses and interviews about race/ethic stereotypes and other prejudices form a necessary part of the work. The professional researcher working with an AR project does not have to be an expert in all these methods, though she must have a solid familiarity with the major alternatives. Rather she must be able to help the group access such research or researchers and guide the process of incorporating these kinds of data in their work.

Real-world human problems are generally multi-dimensional, dynamic, and complex. The conceit of a simplified academic division of labor that tries to treat some issues as appropriate to history, others to political science, others to economics, and so on is useless. Solving real problems in context without oversimplification is a requirement for action research projects.

Action research is not applied research. In applied research, conventional researchers examine a social problem, develop a set of recommendations, and then try to implement a plan of action. They define the problem, they create the recommendations, and they design the intervention, all on the basis of their research and their professional expertise. As Flyvbjerg puts it, they know what "the good life" is (Flyvbjerg, 2001).

In action research, the researcher is an important part of the group but also brings academic and experiential knowledge brought from other projects and from the literature. The local participants are experts in their own lives, problems, and situation, and their knowledge of the details of their problem and the possibilities for action is great. It is up to them to decide what the problem is, that "the good life" would be like, and how it is to be brought into being. Their local knowledge is key, and they will have to implement and live with the consequences of their actions in a way that the professional researcher will not.

Action research is not a social theory but a set of procedures for the deployment of a wide variety of theoretical approaches generated historically in the social sciences. Action researchers are opportunistic in using any and all theoretical approaches that promise to be of help in addressing the problems the stakeholders have identified.

5.3   What Action Research Is

Action research is participatory social research in a variety of senses. It opens up research process to non-professional researchers who are stakeholders in the problem at hand. These non-professional researchers not only provide input for the process but engage in the key decisions about the goals, methods, execution, and interpretation of the results. Action research is about doing research with rather than on people. Because the local stakeholders are directly affected, they have the right collaboratively to guide the process.

Action research is based on the proposition that all significant learning is based on a well-managed interaction between reflection and action. Without reflection, action is incompetently guided but without action, reflection is basically useless. In this way, action research is diametrically opposed to the dominant ideology of the conventional social sciences and conventional applied social research that demands a radical separation of theory and action, of theory and application and accepts the falsehood that it is necessary or possible to theorize without applying the theories in concrete contexts.

Action researchers believe that separating theory and action is the highroad to social science irrelevance and patronizing applied research projects. The irrelevance and public disrepute of much social science is explained by the devaluation of application. It permits academic social researchers to study social problems without taking action and liberates applied researchers from the demands of theoretical sophistication.

Leaving aside the micropolitics of the social scientists, a more epistemologically and methodologically important issue is at stake. Action research is firmly based on scientific methods that require defining problems in an open and clear way, developing a variety of hypothetical explanations for the problem at hand, determining which data are relevant to the analysis of the problem, collecting the data systematically and well, organizing the data, and using the data to test the hypotheses. In the case of action research, as in the case of laboratory sciences, the test of the interpretations is made in context. If the action research–based interpretation of the problem is correct, the actions designed on these interpretations will produce the desired results. If not, the process has to be reiterated, altering hypotheses, collecting different data, working through other interpretations, or all of the above until the outcomes match the expectations.

The contrast between this and social theory developed in the absence of application or applications developed in the absence of theory is stark. Theory without application is mere speculation. Untheorized application is mere guesswork. Of course, separating theory from application makes learning from experience all but impossible.

5.4   The Assumptions Underlying Action Research

The following are some of the key assumptions underlying most action research. Action researchers believe that most important human problems are multi-dimensional, dynamic, and interactive. Therefore, we argue that only multi-dimensional, dynamic, and interactive research strategies can yield meaningful results to such challenges. This involves rejecting the current Fordist division of labor model of the academic world that separates disciplines and expertise into non-interacting silos. These multi-disciplinary successes of the physical, life, and information sciences in recent decades suggest that they have understood these challenges in the same way action researchers do. Not so with the social scientists and humanists.

Action research is based on respect for the knowledge and intelligence of non-academic people. We believe that most people are capable of conducting research, interpreting the results, and designing actions based on these interpretations when the collaborative learning processes are well structured. A corollary of this is that non-academic experience is as important as formal education in conducting efficacious research. The other side of not believing in the knowledge monopoly of academics is for all participants to learn to share their diverse experiences, skills, and hopes and to synthesize this diversity into shared knowledge and plans for action. This is also the foundational belief for democracy. Action research is democracy in action.

5.5   The Origins of Action Research

Action research is not new. It has been around since the beginning of the Western intellectual tradition but it has been increasingly suppressed as capitalism extended its grasp over the global system.

5.5.1   The Philosophical Bases of Action Research

All the key bases for action research are clearly present in the work of Classical Greek thinkers. Aristotle is perhaps the key source of the relevant concepts for action research. His distinctions between kinds of knowing into *epistêmê*, *tekhnê*, and *phrónêsis* has been revisited repeatedly in recent years by Olav Eikeland (2008),[2] Stephen Toulmin (1990), Stephen Toulmin and Björn Gustavsen, Eds. (1996), and Bent Flyvbjerg (*op. cit.*) to show that the contemporary dichotomy of knowledge into theory and application is not only wrong but is a profound dilution of the Aristotelian legacy. *Phrónêsis* is not just an essential ingredient but is also the source of the most valued forms of social knowledge such as clinical knowledge.

A second major ingredient in this philosophical genealogy is the work of the American pragmatist philosophers William James (1948, 1995) and John Dewey (1900, 1902, 1991). Their views about the link between thought and action, collaborative learning, and the testing of ideas in the context of application are a core ingredient in all action research. This connects to the work of Wittgenstein on "language games" (Wittgenstein 1953, Monk 1990), Habermas (1984, 1992) on "ideal speech situations", Gadamer (1982) on hermeneutics, and Rorty (1981) on neo-pragmatism.[3]

Thus there is a very significant philosophical basis for action research, a philosophical basis that most conventional researchers ignore.

5.5.2   The Social Bases of Action Research

Action research has re-emerged at this point in history as a counter-proposal to the hopeless situation of the academic and applied social sciences as currently organized. This problem began with the creation of doctoral programs and professionalization of the social sciences in the latter part of the nineteenth century. Beginning as political economy with Thomas Malthus, Adam Smith, David Ricardo, and Karl Marx, the social sciences were ripped out of this holistic perspective, first by separating history from political economy, then by separating economics out as a discipline and then subsequently by dividing the remaining turf into sociology, psychology, political science, and anthropology. Coinciding with the creation of doctoral programs in the social sciences in the United States from 1880 to 1910 (Cole 2009; Ross 1991; Madoo Lengermann and Niebrugge-Brantley 1998), this resulted in the fragmentation and academicization of social research that endure to this day.

This fragmentation creates professional academic monopolies that served the interests of academic professionals but that shed the holism and reformist intent of political economy, converting academic social science into a non-threatening, for-professionals-only set of activities. Even the ongoing policy relevance of economics has been deeply troubling to economists who continue to privilege theory over practice and who exiled welfare economics and institutional economics on their road to theoretical purity (Furner 1975). In the words of Slaughter and Leslie (1997), academic social scientists have organized themselves into mini-cartels that, unlike the cartels of advanced capitalism, mainly produce and consume their own products (graduate students, research projects, and professional books and articles). This activity becomes entirely auto-poetic.

The loss of integration that came with the dismemberment of political economy means that the contemporary social sciences define research problems in the light of their own theories and methods rather than taking the problems on in their real-world contexts and complexities. Then by treating application as anti-intellectual, they wall themselves off from the recognition that either their theories and methods don't work or matter to most people. In effect, this has "de-socialized" the academic social sciences. A sure way to have a failed academic career in the social sciences is to show an interest in activist research (Greenwood 2008).

Conventional social scientists actively contribute to the maintenance of class relations through education, engaging in the social production of elites and elitism. It is no surprise that academic neo-liberalism, such as "rational choice theory", has become the dominant paradigm in economics, sociology, and political science.

5.5.3 Counter-movements

Historically there have been counter-movements against the hegemony of the abstracted, professionalized, disengaged social sciences. The battle between political economy and neo-classical economics raged for a generation or more (Furner, *op. cit.*) and gave rise to academic purges. Reformist sociology was quickly purged of its reformers like Jane Addams (Magoo Lengermann and Niebrugge-Brantley, *op. cit.*). The anti-Jim Crow, anti-American Indian genocide, and anti-immigration quota commitments of the American anthropologists were quickly muted. By the time of the McCarthy era, most academic social scientists knew how to steer as clear of social reform work and engaged in self-censorship to be sure they stayed socially irrelevant (Price 2004).

Despite this many counter-movements have come and gone. Institutional economics with people like Thorstein Veblen and Clarence Ayres had its brief moment. The Human Relations movement in sociology had a similar rise and fall. The founding of the Society for Applied Anthropology signaled a rejection of academic business as usual in American anthropology but it too was quickly domesticated. And history has repeated itself with science and technology studies, feminism, and ethics studies, each starting as a reform movement and then being "disciplined" into conventional socially distanced academic activities (e.g. Messer-Davidow 2002).

Outside of academia, some parts of NGO world have attempted to challenge the hegemony of authoritarian neo-liberal international development agencies like the World Bank, the IMF, and USAID with some small successes. As yet they have not mounted an effective challenge to the hegemony of these organizations that directly promote global capitalist interests. Liberation movements in the global "South", including Catholic Action, Marxist-inspired organizing, adult education, organizations inspired by liberation theology, and even some evangelical groups have challenged business-as-usual in international development and developed significant momentum for some periods (Freire 1970; Fals Borda and Rahman 1991; Horton 1990; Horton and Freire 1990; Hisdale et al. 1995; Belenky et al. 1997a, b; Park et al. 1993).

5.6   How to do Action Research?

A synthetic presentation of how to do action research is misleading because the sequences, issues to be engaged, and contextual conditions always affect what is possible and what is done. So the following is really an abstract model of processes that, on the ground, often look quite different but that do not violate the general principles articulated here.

AR begins with either the formation of a group of interested stakeholders with a shared problem or with joining an already-existing group of the interested stakeholders. How this happens varies greatly. Sometimes stakeholders seek out an action researcher for assistance. Sometimes action researchers have garnered some resources that could help solve an important problem and go out to create a collaborative group. Once the group is formed, the participants, after getting to know each other and after time spent learning about the pressing problems, engage in a collaborative problem selection process. Often there are more problems than resources (in time and money) than can be dealt with and the collaborative group has to engage in a process of prioritizing among issues.

When the problem to be dealt with has been selected, a process of working through as many possible explanations for the existence and persistence of the problem are examined. This often involves the professional researcher bringing in what is known about the problem from the published literature and from her experience and lengthy discussions among the other stakeholders about their experiences of and understandings of the problem.

From this emerges a set of research requirements that the group must meet in order to deal effectively with the problem. This requires division of research labor among all the stakeholders and the provision of training for the interested stakeholders so they can engage more effectively in the research processes. The research process is planned and then a period of collaborative research ensues in which some people work individually, others in teams, and there are meetings to share research problems and preliminary results.

As the initial research phase closes, the group engages in a comprehensive sharing of results and subjects all the work to critique and interpretation. Problems, oversights, and findings are all evaluated and gradually a vision of the obstacles standing in the way of solving the problems emerges. At this point, the group engages in collaborative design of actions to remedy the problem and a specific action plan for undertaking the change process. From this, the group becomes an action team, applying their action designs to the problem and gathering information about the results. If the outcomes are not what were expected or if new obstacles emerge, the group recycles the action design and implementation process until a better link between the actions and positive outcomes is achieved. This necessarily involves data collection about the results and a systematic and open analysis of the effectiveness of the actions backed up with data about the results capable of convincing third parties of the credibility of the claims.

Often this process is diagramed as an ascending spiral rather than as a linear plan because cycles of reflection and action often cause modification in the initial problem formulations, interpretations, strategies for action and the group may move various times through problem formulation, research, action design, action, and evaluation (Reason, ed. 1988, 1994).

5.6.1   Role of Professional Researcher

One of the unique features of action research is the role of the professional researcher. Research training, skills in methods, knowledge of theory, experience in research are all essential to AR processes. The professional researcher needs to be a well-trained professional social researcher with a broad multi-disciplinary background. But the action researcher is not the solo researcher who does research on and for others. Rather she is a facilitator of group processes leading to the creation of more effective learning arenas for the other stakeholders and herself. She is a teacher but also a learner from the store of experience and judgment of the other stakeholders. She is a facilitator but also a collaborator who participates in the research process directly and also coaches the other researchers.

As an experienced social scientist, the action researcher already has a good deal of training and experience in organizing data, formulating interpretations, and synthetically writing about what is being learned. But, while it is often too tempting, she does not do all the writing or dominate the representation of the work. She is expected to serve as an assistant in developing texts and presentations based on the shared experiences but also to help others learn these skills in the course of a project. Where the educational level of the other stakeholders is very modest, the requirements that the professional researcher do the writing are greater. But action researchers must always remain alert to the way that rendering the projects in writing can co-opt the voices and knowledge of others. Thus, even as a solo writer, action researchers are expected to take the writing to the collaborators and explain what is being said about the project and make the modifications they deem relevant.

Action researchers also do write for other action researchers. Reflecting on what they have learned in carrying out projects and generating ideas they want to share to help improve the practice of other action researchers is not just a legitimate but necessary activity. However, this is a separate intellectual task from supporting an action research project team.

Thus the action researcher is not the boss, not the solo intellectual, not the team leader but a specialized team member who brings training, techniques, theories, and methods as needed in support of the group's efforts and who facilitates the collaborative learning process in the group.

5.7   Examples of Action Research Projects

Despite the convenience of imagining that all AR, or any other social research projects for that matter, develop according to an ideal plan, this is never the case. AR project development is as diverse as the projects themselves and the AR strategies used. What they have in common is a commitment to democratizing the research process and creating outcomes that the stakeholders see as positive. In that spirit, I will give two examples from my own experience to emphasize how different projects can be in terms of starting points, goals, kinds of collaborators, and institutional settings.

5.7.1 The Mondragón Project

This project took place within the Central Services department of what was then the FAGOR Cooperative Group in the labor-managed cooperatives of Mondragón (now linked in an overall group, the Mondragón cooperatives. *Mondragón: Humanity at Work* is the current name of the group. Founded in 1956, these labor-managed cooperatives are the most successful in the world, now employing over 75,000 worker-owners in

242 companies on five continents. Because they are so unusual in being competitively successful labor-managed organizations they are the subject of many studies (Thomas and Logan 1982; Whyte and Whyte 1991). One of the researchers, a famous professor of industrial and labor relations at Cornell University, William Foote Whyte, was conducting a study of them (Whyte and Whyte, *op. cit.*) and, in the process, he offered a feedback seminar to his hosts with his critiques of their operations. The head of human resources for the FAGOR Group thanked him for his critique and then asked him how he intended to help them solve the problems.

Whyte knew that I was an expert on the Spanish Basque Country and involved me in what became a funded project with the cooperative members. Since they had immense experience in collaborative group problem-solving, I decided that any attempt to solve their problems should be built on those practices and be consistent with the cooperative approach to collaborative management. To that end, we convened the research stakeholders they chose and I facilitated a long series of seminars aimed at finding out what problems they had that they wanted to solve and then figuring out what research techniques and processes they needed to learn and apply to develop solutions.

Over the course of what became a three-year AR project, they determined their core concern was that they were adding so many new members who were recruited because of well-paying, stable jobs in a good work environment. They worried that these recruits did not share the democratic values of the cooperatives. The research we did subjected this view to examination and we discovered that it was quite wrong. They were right that people were recruited by the good jobs and conditions but they learned that new workers soon came to value the cooperative approach highly. The dissatisfaction and alienation these worker-managers experienced stemmed rather from the hierarchical, authoritarian ways human resources were handled. In effect, the new recruits were disappointed about the failure of the management of the cooperatives to live up to cooperative values. The result was the need for major changes in the mode of operation of the human resources organizations and these changes were put into practice. In addition, the members of the research group also became an internal consulting organization for cooperatives having human resource problems and used AR as the way of working on change projects. Finally, we together wrote two books about this work (Greenwood et al. 1990, 1992), and the Spanish-language version was used as part of the training program for new hires.

From beginning to end, I combined group process facilitation and consulting on research methods. However, I was also a teacher of methods and social science interpretation and writing, doing some of my own writing but mainly helping the cooperative members develop their own research and writing skills.

5.7.2 Ford Canal Corridor Initiative

The Housing and Urban Development authority of the US government gave out a significant amount of money to try to re-develop the old barge canal system in New York State for the purpose of creating tourism and related employment opportunities. The barge canal system had been the principal transport network for manufacturing in the period before the railroads and interstate highway system and many small industrial towns sprang up and prospered along the canals. However, when that transportation system lapsed, the towns fell into a long cycle of de-industrialization and population loss. The concept of the HUD Canal Corridor Initiative was to restructure the canals as a tourism asset and attempt to create new economic opportunities for these small towns.

After a period of grants, HUD put out a call for proposals to evaluate the results and a group of sociologists, planners, and economists from Cornell University submitted a proposal. The general proposal was of interest to HUD but they insisted on adding an action research dimension to it. The assembled program team had no competence in AR and thus contacted me and Frank Barry, Senior Extension Associate, in the Family Life Development Center at Cornell to help. In return for training the group in AR, we insisted on doing small AR projects in two canal corridor communities on the subject of community development.

In both communities, Frank Barry had prior contacts regarding youth programs. Since opportunities for youth and preparation for adult careers were burning issues in both communities, coalitions of adults and authorities (mayor, superintendent of schools, bankers, teachers, local religious authorities) all had interests in the development of opportunities for youth. We began with these coalitions and engaged in a process by which they expanded the participation in their coalition to include all the major categories of stakeholders in their communities, including the youth themselves. Each of these groups developed a focal issue for an AR activity and we then convened a two-day search conference (a participatory strategic planning process, see Greenwood and Levin, *op. cit.*) in which they developed their shared history of these problems, their understandings of the assets they have and the obstacles they face, and developed action plans in a variety of areas related to youth opportunities.

We provided ongoing support for these action teams during a year's time and then reconvened the two community groups together for a process of sharing their progress and developing their ongoing plans. In neither community were the results revolutionary but both communities developed more collaborative capacities to work on community projects and better linkages with the various state and other funders to help them push forward their community development plans. One of the communities succeeded in getting additional competitive funding for ongoing efforts.[4]

5.8   Varieties of Action Research

There are as many varieties of AR as there are visions of social change and the ideal democratic society. Morten Levin and I have documented the varieties in greater detail in Greenwood and Levin (2007).

Many action researchers take a reformist approach to social change. That is to say, they are critical of existing social arrangements but they are not revolutionaries. They believe in the possibilities of meaningful reform within the structure of current global capitalism. They are not naive and believe that, even if the larger problems of inequality and exploitation cannot be overcome easily, a great deal can be done to improve the quality of life, work, and communities within the existing structures. Examples of these approaches would be Whyte and Whyte (*op. cit.*), Greenwood et al. (*op. cit.*), Reason (*op. cit.*), Heron (1996), and Flood and Romm (1996).

Other action researchers take a more liberationist approach to these issues. While generally not being declared revolutionaries, they take a more directly confrontational approach to power and exploitation and believe in using AR to build the capacities of the oppressed to confront power successfully. Examples are Freire (*op. cit.*), Fals Borda and Rahman (*op. cit.*), Hall (1975), Hall et al. (1982), Horton (*op. cit.*), Hisdale et al. (*op. cit*).

Still others see AR as working as much within the stakeholders as in the larger society. They emphasize the development of psychodynamic approaches to non-defensiveness, the ability to confront power more directly, and learning to work in groups by both leading and supporting leaders in a variety of ways. Their assumption is that more healthy and effective individuals together can bring about significant changes in their own environments and ultimately in society as well. Examples are Argyris (1974, 1980, 1985, 1993), Argyris et al. (1985), Argyris and Schön (1978), Schön (1983, 1987), and Schön et al. (1991), Belenky et al. (*op. cit.*), Hirschhorn (1990, 1998).

Another group of action researchers engages problems at the system level. From their perspective, many of the defects of our society come from larger-scale system processes and need to be addressed systemically. Thus they build large-scale regional and national programs that integrate work across many local sites in larger development coalitions that share their learning and strategies and together attempt to move the larger system in a more democratic direction (Gustavsen 1985, 1992; Levin 1984, 1993, 1994; Flood and Romm, *op. cit.*). There are many more varieties of AR but this inventory points out how very different the practices and strategies can be.

5.9   Visions of Authority in AR

One of the significant ways practitioners of AR differ is in their analytical view and attitude toward authority. These views range from the an almost anarchist belief that all authority relations are an obstruction to human development to a therapeutic view of the AR practitioner as a strong leader who liberates human potential for important change processes to begin. And all the positions in between—facilitator models, team-based cooperation, dialogue leaders, activist organizer models—are represented in the literature (Reason and Bradbury, Eds.). These are differences that make a difference because they affect the way AR processes are initiated, how the collaborators are treated, what kinds of group processes are accepted, what kinds of changes are considered to be significant or worthwhile, and how success is measured. No one of these positions is correct. Knowing the difference matters since at the very least, an AR practitioner has the obligation to understand her own theories and practices of change and to have clear ethical standards that guide her conduct over the course of projects.

Tables 5.1 and 5.2 attempt to capture the implications of some of the key differences in approaches for the practice of action research.

Regarding the vision of social change that inspires the action research project, the tables point to four different views of the key locus of significant democratic social change. Each of these visions tends to correspond in certain ways with the four approaches to action research intervention I have laid out. The reformist/collaborative approach can use diverse approaches but not the "organizer" view that assumes that the outside organizer knows better what local stakeholders need. The systems interventionist practitioner must have a prior notion of the system of which the problem is a part and thus shares some of the vision of the organizer but also sees dealing with particular key processes in the system itself as a central point. The psychodynamically inclined practitioner tends to focus on dialogue and individual psychodynamics as the keys to producing change in a system.

Table 5.1 Vision of social change

Intervention approach	Reformist/ collaborative	Liberationist/ confrontational	Systems practice	Psychodynamic interventionist
Facilitator/ dialogue leader	X			X
Systems interventionist	X		X	
Organizer		X	X	
Psychodynamic	X			X

Table 5.2 Approach to participation

Intervention approach	Input	Consultation	Collaboration	Self-managing group
Facilitator/dialogue leader			X	X
Systems interventionist	X	X	X	
Organizer	X	X		
Psychodynamic	X			X

Regarding participation, these approaches also differ a good deal. The facilitator/dialogue leader emphasizes collaboration and the group's skills at self-management. The systems interventionist emphasizes input, consultation, and collaboration but is somewhat less focused on creating self-managing groups and more on creating healthier systems processes. The organizer wants input and consultation, but because organizers already have a vision of the changes needed, they are less interested in collaboration than in group discipline and not very committed to the creation of self-managing groups that may depart from the organizer's strategy. The psychodynamic approach stresses extensive personal input and the creation of dynamics that lead to self-managing groups that are characterized by healthy psychological attitudes and processes.

5.10 Particular Problems Encountered in AR

While this does not distinguish AR from other forms of social research and intervention, it is important to acknowledge that no AR project ever is perfect. All projects fall short of perfection and participatory processes can always be enhanced and deepened, no matter how successful they are. Since conventional research also rarely goes according to the ideal plan, it is important to explain why this is particularly important in AR. People engaging in AR projects are committed to democratizing social situations, to the ethical treatment of collaborators, and to the possibility of major social improvements in concert with strong ethical beliefs. Given that, the stakes in an AR project are very high for the participants. Not having a project work perfectly and according to plan; having conflicts break out occasionally in the group; and not getting everyone to participate fully can feel like failure in more than a research project—it can feel like failure in democracy and a lack of integrity. This matters because no AR project is perfect and novice practitioners may become discouraged early in projects when they do not develop perfectly and they may assume they are doing something wrong. AR projects succeed incrementally, on a day-by-day basis, and with backward and forward movements throughout.

Many AR projects never realize the full potential of AR. Sometimes the conditions simply are not suitable. It may be that political conditions militate against it. It may be that there simply are not enough financial resources to carry on. Or a group may simply run out of energy before achieving all of their goals. This is common in AR work but many partially realized AR projects do some real good. People learn new skills, gain new perspectives on important problems, solve some but not all the problems they face. These are all real accomplishments and the incompleteness of the project should not cause the participants to lose sight of what they have accomplished.

The professional authority and professional respect are very much in play in AR projects. Particularly in the early phases, the local stakeholders' confidence that the professional researcher knows what she is doing and has plan for the group is important in developing the kind of group dynamic to enable participants to take control of a developmental process themselves. And yet the professional needs to operate more like a midwife than like a surgeon. The professional needs to be attentive to training other participants in the approaches, in ceding or in demanding that authority be shared (along with responsibility), and in not dominating the air space and the communication about the project. Since conventional professionals are trained to want a high degree of authority, to see themselves in a specialist and technical role that makes them superior to the other participants in certain ways, learning to be professional and not to be domineering requires both practice and self-discipline. Unless the professional actively presses in this direction, the default result is that people defer to the professional, eventually get alienated from the process, and withdraw their interest and participation.

The status of the professional also involves intellectual property issues. The currency of professional activity is a combination of written products and compensation that are the supposed requirements of a professional role. The question of intellectual property is vexed in AR because the intellectual property created by the participants is a joint creation that would not have happened easily without professional facilitation. At the same time, many participants are not interested in or do not feel able to write about what they have done and there is a strong tendency for the professional to be expected to do all the writing.

This is a complex matter. The professional is an experienced writer, someone who has learned how to take a variety of materials, synthesize them and put them into narrative form. The professional also may be learning things in the process that are of interest to other professional colleagues but not to other participants in the project. The best solutions are to be open about the intellectual property issues, to work out agreements about who writes and speaks with whom about what, and what rights of review by the stakeholders and the professional exist. Often this works out reasonably well by means of writing some things together or in mutual consultation for the project and by the professional writing other things for professional colleagues, things not so much of interest to the other stakeholders but over which they have some say in deciding if they have been fairly described. In my own personal experience, one of my richest AR experience came from an ambitious and extended project of writing the results up with the other participants. They varied a good deal in their comfort with writing but the time spent working on drafts, debating analyses, and the rest of the discipline that goes with writing was described by one member of the group as the richest learning experience in the whole project.

Since AR is based on both a respect for diversity and a belief that the diversity of experience, perspective, and capability is one of the most important resources an AR group has, dealing with diversity positively is essential to AR projects. However, dealing with diversity by avoiding any conflict, reconciling all differences by lowest common denominator solutions, and by being politically correct rather than honest can undermine an AR project entirely. All stakeholders have a right to articulate their views, to debate with others, and to disagree when they sincerely don't agree. But AR projects proceed by leaving contentious issues that cannot be resolved aside and concentrating on those actions that people can agree to take forward. Sometimes an experience of success with a few issues can make it possible for groups to go back to more divisive issues with new energy and confront those as well.

Paolo Freire's goal of "speaking the truth to power" sounds wonderful but needs to be thought through carefully. Sometimes doing so can bring the immediate destruction of a group of stakeholders. In such cases they should avoid confrontations, at least until they have become well enough organized and supported to be able to deal with a direct confrontation. In any case, one rule of AR is not to take risks for other people. Therefore, taking actions in risky situations must be analyzed carefully in the group. Here the facilitator has important responsibilities because some group dynamics lead people to be silent in the face of power and that has to be confronted. On the other hand, another kind of group dynamic can lead people collectively to feel obligated to take risky actions that as individuals they would not take. This is called "risky shift" in social psychology (Wallach et al. 1962). The professional facilitator has a clear obligation also to be alert to this dynamic and to discourage the group from taking more risks than the members who make it up feel comfortable with.

Action research directly confronts the academic social scientists and is obligated to "speak the truth to" academic social scientists about their complicity in the *status quo* through their face-saving distinction between rigor and relevance, between objectivity and engagement, between analyzing and acting. Everything in AR militates against the validity of these distinctions and thus rejects the bedrock of the practices of the abstracted academic social sciences. To the extent one is to be an action researcher, one must be ready to confront the academic establishment and to face the hostility that unmasking the convenient ways the academic social sciences evade action and social responsibility necessarily creates. Conventional researchers oppose AR because AR questions their right to do what they do and questions the reward structures that support their behavior.

Notes

1. For general references on action research see Greenwood and Levin (2007), Stringer (2004), Stringer (2007), and Reason and Bradbury, Eds. (2007).
2. Of these, Eikeland's is the most reliable and fully explained resource and also has the virtue of being written by a philosopher with a quarter century of action research experience.
3. For a critical review of pragmatism, see Diggins (1994).
4. For an analysis of this project, see Schafft and Greenwood (2003).

References

Argyris, C. (1974). *Theory in Practice*. San Francisco: Jossey-Bass.

Argyris, C. (1980). *Inner Contradictions of Rigorous Research*. New York: Academic Press.

Argyris, C. (1985). *Strategy, Change, and Defensive Routines*. Boston: Pitman.

Argyris, C. (1993). *On Organizational Learning*. Cambridge, MA: Blackwell.

Argyris, C., & Schön, D. A. (1978). *Organizational Learning II Theory, Method, and Practice*. New York: Addison Wesley Publishing Company.

Argyris, C., Putnam, R., & McClain Smith, D. (1985). *Action Science: Concepts, Methods, and Skills for Research and Intervention*. San Francisco: Jossey-Bass.

Belenky, M., Bond, L., & Weinstock, J. (1997a). *A Tradition That Has No Name*. New York: Basic Books.

Belenky, M., Clinchy, B., Goldberger, N., & Tarule, J. (1997b). *Women's Ways of Knowing*. New York: Basic Books.

Cole, J. (2009). *The Great American University*. New York: Public Affairs.

Dewey, J. (1900). *The School and Society*. Chicago: University of Chicago Press.

Dewey, J. (1902). *The Child and the Curriculum*. Chicago: University of Chicago Press.

Dewey, J. (1991). *The Public and its Problems*. Athens: Ohio University Press. (Original work published 1927).

Diggins, J. (1994). *The Promise of Pragmatism*. Chicago: University of Chicago Press.

Eikeland, O. (2008). *The Ways of Aristotle: Aristotelian Phrónêsis, Aristotelian Philosophy of Dialogue, and Action Research*. Bern: Peter Lang.

Fals Borda, O., & Rahman, M. A. (Eds.). (1991). *Action and Knowledge: Breaking the Monopoly with Participatory Action Research*. New York: Apex.

Flood, R., & Romm, N. (1996). *Diversity Management: Triple-loop Learning*. Chichester: Wiley.

Flyvbjerg, B. (2001). *Making Social Science Matter: Why Social Inquiry Fails and How it Can Succeed Again*. London: Cambridge University Press.

Freire, P. (1970). *The Pedagogy of the Oppressed*. New York: Herder & Herder.

Furner, M. (1975). *Advocacy and Objectivity: A Crisis in the Professionalization of American Social Science*. Lexington: University of Kentucky Press.

Gadamer, H. G. (1982). *Truth and Method* (2nd ed.). New York: Crossroads.

Greenwood, D. (2008). Theoretical Research, Applied Research, and Action Research: The Deinstitutionalization of Activist Research. In C. R. Hale (Ed.), *Engaging Contradictions: Theory, Politics, and Methods of Activist Scholarship* (pp. 319–340). Berkeley: Global, Area, and International Archive, University of California Press.

Greenwood, D. J., & Levin, M. (2007). *Introduction to Action Research: Social Research for Social Change* (2nd ed.). Thousand Oaks: Sage.

Greenwood, D., et al. (1990). *Culturas de Fagor: Estudio antropológico de las cooperativas de Mondragón*. San Sebastian: Editorial Txertoa.

Greenwood, D., et al. (1992). *Industrial Democracy as Process: Participatory Action Research in the Fagor Cooperative Group of Mondragón*. Assen-Maastricht: Van Gorcum.

Gustavsen, B. (1985). Work Place Reform and Democratic Dialogue. *Economic and Industrial Democracy, 6*, 461–479.

Gustavsen, B. (1992). *Dialogue and Development*. Assen-Maastricht: Van Gorcum.

Habermas, J. (1984). *The Theory of Communicative Action: Reason and the Rationality of Society*. Boston: Beacon Press.

Habermas, J. (1992). *Moral Conciousness and Communicative Action*. Cambridge: MIT Press.

Hall, B. (1975). Participatory Research: An Approach for Change. *Convergence, 8*(2), 24–32.

Hall, B., Gillette, A., & Tandon, R. (Eds.). (1982). *Creating Knowledge: A Monopoly? Participatory Research in Development*. New Delhi: Society for Participatory Research in Asia.

Heron, J. (1996). *Co-operative Inquiry: Research into the Human Condition*. London: Sage.

Hirschhorn, L. (1990). *The Workplace Within: Psychodynamics of Organizational Life*. Cambridge: MIT Press.

Hirschhorn, L. (1998). *Reworking Authority: Leading and Following in a Post-Modern Organization*. Cambridge: MIT Press.

Hisdale, M. A., Lewis, H., & Waller, S. (1995). *It Comes from the People*. Philadelphia: Temple University Press.

Horton, M., & Freire, P. (1990). *We Make the Road by Walking: Conversations on Education and Social Change*. Philadelphia: Temple University Press.

Horton, M. (with Kohl, J., & Kohl, H.). (1990). *The Long Haul: An Autobiography*. New York: Doubleday.

James, W. (1948). *Essays in Pragmatism*. New York: Hafner.

James, W. (1995). *Essays in Radical Empiricism*. Lincoln: University of Nebraska Press.

Levin, M. (1984). Worker Participation in the Design of New Technology. In T. Martin (Ed.), *Design of Work in Automated Manufacturing Systems* (pp. 97–103). Oxford: Pergamon Press.

Levin, M. (1993). Creating Networks for Rural Economic Development in Norway. *Human Relations, 46*(2), 193–217.

Levin, M. (1994). Action Research and Critical Systems Thinking – Two Icons Carved out of the Same Log. *Systems Practice, 7*(1), 25–41.

Madoo Lengermann, P., & Niebrugge-Brantley, J. (1998). *The Women Founders: Sociology and Social Theory 1830–1930, A Text/Reader*. Long Grove: Waveland.

Messer-Davidow, E. (2002). *Disciplining Feminism: From Social Activism to Academic Discourse*. Durham: Duke University Press.

Monk, R. (1990). *Ludwig Wittgenstein: The Duty of Genius*. London: Jonathan Cape.

Park, P., Brydon-Miller, M., Hall, B., & Jackson, T. (Eds.). (1993). *Voices of Change: Participatory Research in the United States and Canada*. Westport: Bergin and Garvey.

Price, D. (2004). *Threatening Anthropology. McCarthyism and the FBI's Surveillance of Activist Anthropologists*. Durham: Duke University Press.

Reason, P. (Ed.). (1988). *Human Inquiry in Action*. London: Sage.

Reason, P. (Ed.). (1994). *Participation in Human Inquiry*. London: Sage.

Reason, P., & Bradbury, H. (Eds.). (2007). *Handbook of Action Research* (2nd ed.). London: Sage.

Rorty, R. (1981). *Philosophy and the Mirror of Nature*. Princeton University Press: Princeton.

Ross, D. (1991). *The Origin of American Social Science*. Cambridge: Cambridge University Press.

Schafft, K., & Greenwood, D. (2003). Promises and Dilemmas of Participation: Action Research, Search Conference Methodology, and Community Development. *Journal of the Community Development Society, 34*(1), 18–35.

Schön, D. (1983). *The Reflective Practitioner*. New York: Basic Books.

Schön, D. (1987). *Educating the Reflective Practitioner*. San Francisco: Jossey-Bass.

Schön, D. (Ed.). (1991). *The Reflective Turn*. New York: Teachers College Press.

Slaughter, S., & Leslie, L. (1997). *Academic Capitalism: Politics, Policies and the Entreprenurial University*. Baltimore: Johns Hopkins University Press.

Stringer, E. (2004). *Action Research in Education*. Thousand Oaks: Sage.

Stringer, E. (2007). *Action Research* (3rd ed.). Thousand Oaks: Sage.

Thomas, H., & Logan, C. (1982). *Mondragón: An Economic Analysis*. London: Allen and Unwin.

Toulmin, S. (1990). *Cosmopolis: The Hidden Agenda of Modernity*. Chicago: University of Chicago Press.

Toulmin, S., & Gustavsen, B. (Eds.). (1996). *Beyond Theory*. Amsterdam/Philadelphia: John Benjamins.

Wallach, M. A., Kogan, N., & Burt, R. B. (1962). Group Influence on Individual Risk Taking. *Journal of Abnormal Social Psychology, 65*, 75–86.

Whyte, W. F., & Whyte, K. K. (1991). *Making Mondragón* (2nd ed.). Ithaca: ILR.

Wittgenstein, L. (1953). *Philosophical Investigations*. London: Blackwell.

6

Ethnography and the Management of Organisations

Tony Watson

6.1 Introduction

Ethnography is not a new activity. However, there is currently in organisation studies a new interest in ethnography as scholars increasingly recognise the value of research writing that takes readers deeply inside organisations. Also new, I feel, is the realisation that such research may be the only way to increase our understanding of those people whose work is critical to every organisation: the managers. And how, I wondered, might I explain to readers of the present book the nature of ethnography and demonstrate its potential for getting close to the work and lives of managers. I could run through the few existing ethnographic writings on managerial work, picking out features which might inspire and inform people considering doing ethnography. But there exists so little ethnographic work on managerial work to serve this purpose. So, alternatively,

T. Watson (✉)
Nottingham University, Nottingham, UK

© The Author(s) 2018
M. Ciesielska, D. Jemielniak (eds.), *Qualitative Methodologies in Organization Studies*,
https://doi.org/10.1007/978-3-319-65217-7_6

I could operate in a traditional textbook manner and present in a magisterial voice lists of do's and don'ts that might help potential ethnographers. But this did not feel right. It would not help me to convey what I might call the essential spirit of ethnographic work. But, then again, how could one write about something as ethereal as an 'essential spirit'? I answered this question by deciding to write a chapter looking at some of my own ethnographic endeavours in what I take to be the style of a good ethnographer.

If I were to attempt to write in an 'ethnographic spirit', I said to myself, I could treat my years of experience of this broad style of research as an account of the 'field' of ethnographically inclined research in managerial settings. The chapter would thus be one of those 'tales of the field' that John Van Maanen (1988) writes about when looking at the nature of ethnography, but with the 'field' here being ethnographic work itself. And this is what I decided to do. I recognised that such a venture would not be as easy as it might at first seem. The main challenge was one of pulling off the trick of being simultaneously the writer of a chapter and the subject of it. What do I mean about being the subject of one's own writing? Well, I do not mean engaging in what some writers call 'autoethnography'. My own biography is not in itself likely to be of interest to readers. What might be of interest, however, is an account of the learning process involved in making ethnographies. A classic notion that ethnographers have used over the years, since its introduction in the writing of Geer et al. (1968), is that of the people being studied (medical students in this case) *learning the ropes* of the occupation or organisation in which they are involved. What this chapter does, then, is to share with readers 'the ropes' of ethnographic endeavour in managerial contexts, as I have learned them. To put this another way, everything that I have to say about ethnography comes from what I have learned about succeeding 'in the field' in my various attempts to produce effective and interesting accounts of 'how things work' in organisations, occupations and workplaces. Having said this, I feel it is important to establish just what I mean about a key characteristic of ethnography: the research aspiration to produce *truthful* accounts of 'how things work' in the social world.

6.2   Researching 'How Things Work' in Organisations

Research students are typically encouraged to be clear about their research questions when they embark on a study. Although it is often difficult to be explicit about the research questions behind ethnographic enterprises, it is helpful to try to be explicit about just what it is one is trying to discover or understand better than previously. Let me illustrate this from some early experiences. After graduating in sociology (with 'industrial sociology' as my special interest) I decided that I would like a career as an academic industrial sociologist. And I recognised that to do this I would need to acquire a research degree. There were several factors that discouraged me from becoming a full-time research student, but significant among these factors was the type of broad research questions I had in mind. These questions were shaped by the discussions I had had with my father over the years about the way relationships and activities were managed in the factory where he worked as a spray painter. Why, in particular, were there so many tensions and time-wasting conflicts between managers and workers and between some managers and other managers in that factory?

I learned from friends in other workplaces and from experiences in the organisations where I worked in vacations that this was not abnormal. In my degree work I inevitably came across plenty of studies offering concepts, insights and theories which were relevant to my type of question. But I found very little research which looked at the managerial processes that play a key role in all this. Outstanding, however, was an ethnography in which the researcher actually became an industrial manager and was able to analyse managerial work *from the inside*. This was Melville Dalton's ethnographic study *Men Who Manage* (1939). But I could see no way in which I, as a doctoral student based in a university sociology department, might tackle the sort of questions which Dalton's book had inspired in me—to put it simply, questions about how things generally tend 'to work' among managers and others in work organisations. My industrial sociology tutor then asked me one day when I was reflecting on my future career, 'Why not seek a trainee managerial role in a large work organisation, register for a part-time research degree and, at the same time, earn a reasonable living whilst you investigate the issues which interest you. And you would end up with a qualification which can launch you into an academic job'.

I did indeed enter a junior management position in one of the world's biggest and most successful aerospace companies. And I not only obtained a research degree but also achieved a professional management qualification—this combination of qualifications being extremely helpful in obtaining an industrial sociology lectureship in the fast-growing academic world of business and management studies. This is all very well. I've told a nice story of career success have I not? And I would be very pleased if I were able to say that I had become an ethnographer through my participant observation investigation. This was indeed true in a *de facto* way. I felt I really 'knew the ropes' about managerial careers and how 'things work' in managerial circles. However, at that time, writing in a fully ethnographic style, recounting events as they occurred and reporting in a reflexive manner on the day-to-day politicking which goes on in managerial ranks, was not seen as a good way forward for an aspiring academic. When I published a journal article on the foundry research (Watson 1982) I got away with simply referring to my 'case study'. I did not even mention my participant observation work, let alone use the term 'ethnography'. And in a subsequent research study on the personnel management occupation, emphasis was placed in the doctoral thesis, articles and book on the interview-based material that I had gathered from 100 personnel managers. Little attention was paid to the day-to-day insider experiences and acquired insights from my participant insider experience (in the first place as a junior manager and, in the second place, as a more senior 'industrial relations manager').

So, what sort of thing had I learned in all of this with regard to 'how things work' in the managerial world? To answer this question, I'll look back to my earliest research venture. Here, considerable conflicts were surfacing over the prospective opening of a very large foundry and it became apparent that a failure of the foundry's senior managers to explain, consult and negotiate over aspects of what was to be an enormous change in everyone's life was leading to powerful opposition by practically the whole workforce to a move which was to occur only six months later. Having gained the sponsorship of the corporate Personnel Director, I took on an advisory role with the senior foundry-management team. The team was told that I was 'a qualified industrial sociologist'. And, on the basis of my extensive informal 'networking' across the foundry and a formal workforce survey which I designed and carried out (more to give my arguments credibility with the largely engineering-trained managers than to tell me things I had not discovered as a participant observer), I predicted that there would be a foundry strike well before anyone moved into the new 'casting facility'. There followed hours of argument and debate and one or two angry attacks on 'graduate know-alls'. The head of the foundry even called me in to sack me from the company one day, only to be told that I was on the Personnel HQ payroll rather than his. I was clearly learning very quickly 'the ropes' of participant observation research in management settings.

Essential to understanding these 'ropes' was the recognition that engagement in managerial politics, whether one likes it or not, is vital if the researcher is going to learn anything significant about the running of an organisation. There is no avoiding the necessity of researchers having to manage both 'friends' and 'enemies'. It is an element of 'how things work' in organisational ethnographic investigations. But what about bigger questions of 'how things work' in managerially led organisational change processes more broadly? Here theory and concepts have to come into play. I made central use of a pair of concepts, *orientations to work* and *implicit contracts*, to make sense of the situation in which the senior managers, with two exceptions, were highly committed to personal upward mobility in the company. Very clearly, these men understood that the company, for whom the new foundry was of considerable strategic significance, would reward them well in career terms if they were to succeed with this massive venture. The managers were proud of the scale, design and 'leading edge' nature of what was going to be the world's largest and most advanced 'precision casting facility'. That, at the time of my intervention in the management team, that the bulk of the workforce (middle managers to yard staff, skilled men and women to clerical workers and production engineers) were complaining with increasing bitterness about such matters as the lack of windows in the building, the banning of tea-making on the shop-floor and, very significantly, the attitude of the senior managers that they 'knew best' when it came to the operation of steel foundries. This, I argued in my research writing, reflects the very different class or life-chance situation of non-senior-manager employees in industrial organisations, compared to the situation of senior managers. The former's implicit contract with the company was not centred on rapid upward career mobility. Succeeding with the world-class 'casting facility' would be recognised and rewarded as a great achievement by the senior managers. But it would, for most staff, involve uncertainty, disruption, paying higher bus fares to get to work and, of great symbolic significance, 'drinking management tea from management vending machines'. And, to focus on just one group of workers, the furnace men (a highly skilled group who 'get very thirsty after a shift') would have no public house to go to after work. There was thus a major gap between the priorities, the implicit contracts and the career expectations of the managerial 'dominant coalition' and the rest of the workforce.

At one Sunday morning 'steering group' meeting (where, interestingly from an anthropological point of view, tweed jackets and flannels were worn instead of the weekday 'senior manager' dark business suits) there was an attempt to put 'workforce complaints' down to the company's mistake of allowing 'a bloody sociologist' into the foundry. My response was to argue that my analysis was 'true' and I nervously suggested that it 'fitted with existing sociological research on change programmes'. Most boldly, I suggested to them that I would be willing to return to work in the Personnel HQ as long as they would promise to invite me back to the foundry to help them out when strike action 'starts to bite just before Christmas'. With this, the meeting was adjourned ('the ladies at home will have Sunday lunch ready'). The next day I was invited to remain in the foundry to devise a formal programme to 'involve' the workforce in the final stages of preparation for the move. I am pleased to report that this did happen and that the foundry did not have a strike. And, although I was encouraged not to write up my research in a fully ethnographic style (a notion I shall I explain later), I had learned the ropes of doing participant observation research among managers. And part of the learning that I have passed on to others over the years is that, if you want to gain research access to managerial goings-on, gain insights into strategic activity and, it has to be said, get 'close to power', then there is no option but to deploy the highest level of social, political and rhetorical skills that one can manage. It cannot succeed without this. And this is something I was highly aware of when, 20 years later, I negotiated a one-year secondment from my business school with the senior management of another large company. This was in order to write the book which became *In Search of Management* (Watson 2001), a study I shall come back to shortly.

6.3   Ethnography and Truth-Telling

You may have noticed that, earlier, I mentioned the research aspiration 'to produce truthful accounts of "how things work" in the social world'. But the notion of truth is an exceedingly difficult one to use. I would have struggled at the time of my early research to fully articulate a philosophically and sociologically sound explanation of what I take 'truth' to be. Yes, I needed to persuade both my career sponsor and the foundry managers that my analysis of the problems in the foundry was a true one. One rhetorical move, offered by me and taken up by my Personnel Director sponsor, was to invoke the notion of professionalism and (social) scientific knowledge. The implication was what I said should be accepted and my advice acted upon because I was a trained and qualified industrial sociologist. Probably much more significant was the threat 'on your heads let it be if you ignore Tony's analysis'. In reality (I nearly said 'in truth'), one cannot predict events like strikes any more than one can simplistically present analyses of workforce attitudes as 'facts'. However, I was very happy to argue that the definition of the situation in the foundry that I was putting forward was a much wiser one to work with than the definition adopted by the majority of the senior managers. This was, to put it simply, 'all these negative statements made by the workforce are the result of foolish rumours; once people actually get to this superb new building they will recognise how much their working lives have improved'. This might, of course, be true. But I believed, and strongly argued, that my analysis was the truer one in the sense that, if one were to act on the basis of *my* definition of the situation, then the new foundry venture was more likely to be successful than if one acted upon the foundry senior manager one.

Although I did not clearly articulate this notion of 'relative truths' at the time, it was implicit in the foundry senior managers' eventual acceptance of such a position. And, of course, to adopt my view (supported, remember, by the politically influential Personnel Director from Company senior management) was to avoid any risk to their ambitious career plans. This point has to be made in order to remind us that different 'points of view' or definitions of the situation in organisations must always be understood in the context of organisational politics and career interests. And, of course, this did not just apply to the managers and the workers. I, as the 'expert researcher', was no disinterested party to what occurred; indeed, my career in the company was 'made' by my role in what became the successful and strike-free move to the new 'precision casting facility'.

Twenty years after these events, I found myself embedded for a year in a large telecoms development and manufacturing organisation as a senior manager and (overt) participant observer. My intention was to write what I hoped to be an important book about managerial work 'from the inside'. And it became clear that it would be academically necessary to deal more directly with the question of the 'truthfulness' of ethnographic writing than it had been back in my aerospace days. But it was not just a matter of what I would say to an academic audience. The managers I worked with as a colleague all knew that I was going to write a book at the end of my secondment to the company. And, one day, several of them (all with engineering and science backgrounds as in the previous business) challenged me on the validity of the book that I would write: 'You promised us at the start that you would change names, job titles and various other things so that no individual quoted or their action described in the book would be recognised by readers of the book. So how can you possibly claim that you are going to tell the truth about the managerial work we do?' It so happened that one of those managers had that very afternoon given me a copy of a management magazine that included a feature on the company. This was a glowing account of the brilliant success in change management that the company was achieving. It was clear who the journalist had interviewed for the article, not just because of the terminology, but because the account provided was one which was impressively career enhancing for two particular managers. And the picture painted was one of harmony across the business, both within management itself and between the company and its employees.

No mention was made in the magazine article of either the enormous tensions between different groups of management in the company or the current trade union veto on key elements of the change programme. My colleagues looked at the document and I commented, 'I know from your faces that you don't think much of the article. But we cannot see it as it altogether untrue'. Where I would question the value of the article, I suggested, would be in terms of how helpful it can be as a guide to what is happening in the business if it were given to someone who was coming to work here. The managers agreed that it would be helpful in some ways, but most definitely not in others. It implied that, as one woman put it, 'life here is all sweetness and light. What could be more misleading than that'. And, I said, when it came to reading my book, 'you will probably say that some parts of it are truer than others. However, what I promise you now is that you will find my book to be a lot truer about how things work in the industrial world than what is in this magazine article'.

What I did not go on to say to the managers I was talking to was that I had now found what I referred to earlier as an epistemologically sound justification for this notion of relative truth. I had always had at the back my mind first-year degree-course learning about Popper's view that science can never lead to the discovery of final and irrefutable truths—all it could do was to improve on the existing knowledge current at any given time (Popper 1959). But towards the end of my year in the telecoms factory, my academic reading focused on (American) Pragmatic Philosophy and the distinctive notion of truth claims that it offered. This, like Popper's writing, suggests that there are no final or conclusive truths to be discovered. Any one piece of knowledge, research writing or teaching may, however, be more 'truthful' than another—in the sense that the knowledge in the 'truer' case could act as a better guide to action in the aspect of the world to which it related than the 'less true' one. At the simplest level of the new ethnographer learning the ropes of their trade, this 'test' of relevance to what people might potentially do in light of the knowledge they are creating, this Pragmatist notion of truth is immensely helpful. And, in the broader context of academic research on organisations and management, it suggests an enormously helpful role for ethnographic style research reports. Insofar as the ethnographer's writing is about what they learned by 'getting close to the action', so we will have alternative learning material for management and business students who are currently so dependent on over-rational and prescriptive management textbooks.

6.4   What Do We Mean by 'Ethnography'?

Readers might have noticed that in this chapter so far, 'ethnography' has not been formally defined. And further, nowhere has the research work described and discussed been presented as 'ethnographic research' or the investigative work been portrayed as 'doing ethnography'. Terms like 'ethnographic work', the 'ethnographic enterprise' and 'ethnographic writing' have been used, however. So, what is going on here? Well, strange as it may seem in a chapter written in a research methods book, I want to argue that to get at the essential qualities of ethnography, it is helpful not to treat it as a research method at all. All the research looked at in preceding paragraphs is centred upon the broad research method of intensive field research and, more particularly, on participant observation. Intensive observation, with varying degrees of active participation in organisational processes, is a *necessary* condition for the production of ethnography. But it is not a *sufficient* condition. This is because ethnography is better understood as a form of writing, rather than as an investigative method. My formal definition of ethnography is *a style of social science writing which draws upon the writer's close observation of and involvement with people in a particular social setting and relates the words spoken and the practices observed or experienced to the overall cultural framework within which they occurred* (Watson 2011, p. 205).

The two main clues to finding the essential qualities of ethnography lie in its origins in anthropology and in the word 'ethnography' itself. Thus, we can say that ethnography serves an anthropological interest in understanding the human as a *cultured being* ('ethno') through writing about them ('graphy') in a manner which provides deep insights into humans' *cultured lives*. To talk of 'cultured lives' in this way means relating the details of the particular events and utterances observed, heard and experienced in the field to a *cultural whole* (Baszanger and Dodier 2004; Watson 2012). Thus, my foundry research account set the particularities of events in the aerospace company in the context of the differing social class imagery of people working there and in the emphasis in the organisational culture on managerial status aspiration. And the account of managerial life in the telecoms business contextualised the orientations of the managers and the events which unfolded in terms of competing managerial discourses which exist across the culture of contemporary work organisations and their management. Similarly, a study of a pub and brewing business located the organisation in the context of the role of pubs in the lives of English people and gave particular attention to the social phenomenon of the 'real ale' movement. And ethnographic work in an English village was set in the context of broader processes of urban–rural shifts over previous decades.

Another very good reason not to treat ethnography as a method is in order to keep open the possibility of using a variety of other research methods to complement the essential intensive fieldwork necessary for an ethnography. Earlier, I mentioned the survey carried out in the foundry project. On reflection, I do not think I would have spent the time on this were it not for its 'political' value in my arguments with senior managers. However, in retrospect, I certainly would have done the set of interviews I carried out with each one of the foundry's senior managers. This was enormously helpful in making sense of many of the events in which I had seen them participating. But most fruitfully, it gave me an opportunity to discuss at length, in private and confidential terms, the reservations about the new foundry which I had begun to infer had developed with two of the managers. Initially, each of these men expressed in technical or business terms their reservations about the change. But very soon, as the conversation developed, each of them turned to their own current work orientation. For very different reasons, neither man wanted further promotion in the company. This meant that the enormous disruption in their working lives that was beginning to occur would not be compensated for by the sort of future career 'beyond the foundry' which excited and motivated their colleagues. Coming to understand these two 'deviant' cases provided an analytically powerful *comparative* boost to the understanding of the orientations of the rest of their colleagues. It is possible that a skilled researcher visiting the foundry to interview managers might have elicited the sort of information that came out of my conversations with people whom I had got to know very well as colleagues. But I very much doubt it.

Interviews (in the sense of formally structured and recorded conversations) carried out by 'embedded' researchers are rather different from standard interviews. In my later major ethnographic study I waited until I had spent six months working alongside managers before I set up formal tape-recorded and structured interviews with 60 of them. The 'added value' of interviewers carried out by participant observers is more than a matter of their creating a higher level of trust between the parties (vital though this is) as Spradley explains in his book *The Ethnographic Interview* (1979). It enabled shared experiences and events to be examined and jointly considered to illustrate and 'fill out' day-to-day conversations. And, time and again, it valuably threw light on events and arguments that I had recorded in my day-to-day field notes. It was quite common for interviewed managers to comment to me, in the words of just one of these people, 'You'd never have got all that stuff out of me if it wasn't that I know you well. And I know you've seen enough of me in action for it to be impossible to bullshit you about what a great manager I am'.

In the same way that there can be a process of mutual reinforcement between the outcomes of interviews and the observations made through organisational participation, it is possible that small surveys, quantitative data analysis and documentary discourse analysis can all be brought into service in the process of preparing for an ethnography. The material produced by the use of these methods does not function as additional 'evidence', so to speak. It has to be woven into the fabric of the piece of ethnographic writing as a whole.

6.5   And Finally: Writing One's Ethnography

The phrase 'preparing for an ethnography' was used above to cover all the investigative work carried out by the researcher in the field (as well, often, in the library and in relevant archives). This utterly is not to play down the significance of research work in the field. Ethnography only comes about, however, when this material is pulled together into a clear and coherent narrative together with appropriate concepts and theories from the social sciences, all of this relating detailed and specific matters to broader social and cultural 'wholes'. To give the degree of clarity and coherence that this requires, it is invaluable for the writer to use writing techniques found in 'creative' writing, and in novels particularly. Given its anthropological roots, ethnography draws on both the humanities and the social sciences. If we see both social science writing and high-quality novels as being concerned to identify and reflect upon truths about how the social world works, then it seems wise to develop a form of writing that brings together the strengths of both of these forms.

In ethnography, science provides research questions, concepts, theories and research techniques while creative writing such as novels provides techniques of narrative-shaping, engaging descriptions of people, places and events, and the presentation of research subjects' own words, thoughts and contributions to dialogues and conversations—conversations with each other and with the researcher. In addition to all of this rather challenging set of requirements, the ethnographic writer needs to build a trusting relationship with readers through taking them along with them in their engagement with the particular social setting that they have researched. This is most effectively done by the researcher writing in a reflexive manner: including themselves in the story, so to speak. This enables the reader to take into account whatever biases, interests, purposes, social skills and general human frailties that the investigator was throwing into the fieldwork mix. A wholly objective research account is never possible but one can get closer to it if the researcher/writer has revealed their hand throughout. I hope that I have done this effectively in the present chapter and that, taking into account my clear research preferences and beliefs, readers decide for themselves whether they wish to engage in ethnographic work—with all its tensions, frustrations, joys and opportunities to say something worthwhile and convincing about how organisational management actually works.

References

Baszanger, I., & Dodier, N. (2004). Ethnography: Relating the Part to the Whole. In D. Silverman (Ed.), *Qualitative Research Theory, Method and Practice* (2nd ed., pp. 9–34). London: Sage.

Geer, B., Haas, J., Vivona, C., Miller, S. J., Woods, C., & Becker, H. S. (1968). Learning the Ropes. In J. Deutscher & J. Thompson (Eds.), *Among the People* (pp. 209–233). New York: Basic Books.

Popper, K. (1959). *The Logic of Scientific Discovery*. London: Hutchinson.

Spradley, J. P. (1979). *The Ethnographic Interview*. Belmont: Wadsworth. (Reissued Long Grove: Waveland Press, 2016).

Van Maanen, J. (1988). *Tales of the Field*. Chicago: Chicago University Press.

Watson, T. J. (1982). Group Ideologies and Organisational Change. *Journal of Management Studies, 19*, 259–275.

Watson, T. J. (2001). *In Search of Management*. London: Cengage. (Originally Routledge 1994).

Watson, T. J. (2011). Ethnography, Reality and Truth: The Vital Need for Studies of 'How Things Work' in Organisations and Management. *Journal of Management Studies, 48*, 202–217.

Watson, T. J. (2012). Making Organizational Ethnography. *Journal of Organizational Ethnography, 1*, 15–22.

7

The Agential Materiality of Storytelling

David Boje and Nazanin Tourani

7.1 Introduction

Walter Benjamin (1963/1968, p. 87) tells us "The art of storytelling is coming to an end because the epic side of truth, wisdom, is dying out." Benjamin lists several contributing trends: the rise of the modern novel, which is not grounded in materiality; information that can be too shallow to be useful; and a decline in workplace circumstances where people can get together to practice the art of storytelling and story listening. Benjamin says Nikolai Leskov was a storyteller who forgoes psychological shading, had experience of a situation, and an ability to communicate it in all its process complexity, including attention to its material vividness and material history embedded in the "inscrutable course of the world" (p. 96). The world does not storytell itself. There is no master storyteller

D. Boje (✉)
New Mexico State University, Las Cruces, NM, USA

N. Tourani
Penn State University at Fayette, Lemont Furnace, PA, USA

© The Author(s) 2018
M. Ciesielska, D. Jemielniak (eds.), *Qualitative Methodologies in Organization Studies*,
https://doi.org/10.1007/978-3-319-65217-7_7

giving us master narratives. This kind of thinking deflects attention from the material practices of storytelling. It also deflects attention away from our own complicity, our own ethical answerability in the material apparatus of storytelling practices.

One has to turn to physics, material sociology, and native people's storytelling to find the storytelling that Benjamin sees as passing from the contemporary world. Indeed, this is the intent of our chapter, to revive the material art of storytelling, grounded in the processes of the material world. Despite the fact that there are examples of "everywhen" storytelling in the west, western narrative traditions focus on linearity. Native (indigenous) scholars are breaking with linearity in a return to nonlinear storytelling. In quantum physics, there is a different break with linearity but it also has something to do with storytelling. Finally, in sociology, there is growing suspicion evidence that something is errant about the social construction paradigm, and some rival paradigm is about to topple it. We bring these three trends together and point out some ways methodology can accommodate more nonlinear approaches to storytelling.

7.2   Textually Mediated Communication

We are telling our stories here with many "texts," as conceptualized by the work of sociologist Dorothy Smith and interpreted by communication scholar Martina H. Myers. Texts are communicative objects that allow us to "map" or do sensemaking with our material realities. Texts can be understood as "verbal routines inscribed in organizations like performance appraisals or job interviews" (Fairhurst and Putnam 2004, p. 8). Map and routines are both linear models in the main. The map is not the territory, and a living story is more directly interpenetrating with the territory of relationships, in what Bakhtin (1993) calls once-occurrent Being-as-event. A great deal of what social actors do is habitual and done without conscious awareness. In order to better understand our own and each other's material realities, we often map texts as subroutines that organize our activities, and thus create a higher level of awareness regarding what are typically ongoing, sometimes habitual, nonconscious social activities and relations.

Furthermore, researchers using this approach seek to discover the *texts* that instigate and advance social interactions (Campbell and Gregor 2004; Smith 1990). Texts are those common bits of knowledge that inform and indirectly motivate individuals in regard to what to do, how to behave, and how to interact. In other words, texts are those things we read without thinking consciously about the formation or activation of these texts most of the times (Campbell and Gregor 2004; Smith 1990, 1999, 2005; Myers 2009).

Holstein and Gubrium (1999) extend this discussion of texts and material realities that Smith and Myers describe by investigating the material objects / realities as texts that we construct and which construct us in describing our realities for ourselves and each other. While Myers and Smith focus on texts as the basis for social interactions, Holstein and Gubrium extend this discussion of texts and textually mediated communication into the realm of identity construction and the construction of self. Thus, various texts will constitute and identify our*selves* in the process of doing this research.

As such, the representational paradigm of narrative, for the most part, has aligned with the social construction paradigm, which according to Bruno Latour (2007, 2005) has no material moorings whatsoever. This is not the implication of Aristotle's (350BCE) *Poetics* (see, Aristotle did, however, assume that narrative must have a beginning, middle, and end; a linear and static way of looking at it (Boje 2008)). In this chapter, we introduce a practical sociomateriality approach to storytelling. Before we get to that topic, let us introduce the storytellers: David Boje and Nazanin Tourani.

Example 7.1 Storytellers

I, David Boje, was born in Spokane, Washington, at Saint Vincent's Hospital to Lorane Joyce and Daniel Quentin Boje. I am the eldest of four siblings, and the first in my family tree to attend college, much less to graduate and earn a Ph.D. Both parents and one younger brother have passed away. Somewhere along the way I lost interest in linearity. It could have been that year in Vietnam, in the undeclared war, or an outright refusal to buy into ways of telling that conform to beginning, middle, and end. My main focus is on storytelling. While I am not indigenous or Native American, on both sides of my family tree there were marriages to princesses of tribes in the Northwestern United States. And so I have Boje relatives living on reservations.

I am Nazanin Tourani. I am born in Tehran, the capital of Iran. My parents are both alive, living in my home country. I have grown up and studied up to master's degree in Iran. My engineering background has educated me **mostly** in linear thinking paradigm while culture and especially indigenous folklore has brought me up in diverse nonlinear perspectives. Time is the most prominent nonlinear perspective I have come up with. Not only finding diversity and relativity of time perception among people, but also first-hand experience of living in two different cultural environments (Iran and U.S.) has improved nonlinear perspective of time in my mind. Now, I am in a Ph.D. program in New Mexico State University, where I first met Dr. Boje. Attending Dr. Boje's classes enriched and improved my nonlinear perspective, let me observe the world in a different way. Dr. Boje is my dissertation chair and mentor in my attempts to study strategic alliances.

Next, we introduce a methodology that combines storytelling with materiality. It is called ontological storytelling.

7.3   Vital Sociomateriality and Ontological Storytelling Methodology

What is vital sociomateriality? Vital materialism suggests human as an agent is a complex and rich composition of vital materials, who acquires his powers from matters; hence, his power is a type of thing-power. There is no difference between human and nonhuman agents, but it disapproves of considering humans as an "ontological center." The anthropocentric narrative of the phenomena is not all-embracing in that it ignores nonhuman agents. The remedy is to interpret a phenomenon considering all human and nonhuman agents/actants. Bennett recommends that we improve our attentiveness to things and their agencies by paying more attention to material vitality. She supposes this approach would further affect our political analysis of and reactions to phenomena. Similar insights appear in a variety of philosophies. For instance, it resembles Buddhism's basic tenets of attentiveness, that is, mindful action in Being in many ways. However, the latter insight mainly addresses morale considerations while the former one is an ontological and epistemological philosophy.

Bennett (2010) in *Vibrant Matter* applies Latour's (2005) Actor-Network-Theory (ANT) in developing onto-stories, stories of the assemblages of stuff one encounters and orders in their daily life. Bennett makes a claim similar to Cajete's (2000) Native Science, in saying that matter holds vitality, including an "agentic capacity" (Bennett 2010, p. xii). Such capacity is obvious in the way chemicals affect us. Chemicals in our bodies, as matter instances, influence our behavioral and physical functions. Bennett brings in the concept of nonhuman agents by emphasizing on "material efficacy" (Bennett 2010, p. 9).

By ontological storytelling, we mean a relational process approach. The expert relational process of practiced understanding without worrying about the past or future is also a theme in Dreyfus and Dreyfus (1986). This is an interesting narrative, but I think that Dreyfus and Dreyfus have taken us part of the way down this line in their book "*Mind over Machine: the power of human intuition and expertise in the era of the computer*" (Dreyfus and Dreyfus 1986, p. 30).

Matt brings together human and nonhuman agents contributing to his Taekwondo practice in his ontological story. Following Latour (2005), we storytell in a sociomaterial world, from different times, places, and optics. We focus our examples in this chapter on the storytelling process in relation to living materiality. We will now turn to quantum physics, then to materialist sociology, and finally to indigenous peoples "more materialist" storytelling for new directions in the arts of storytelling.

7.4   Quantum Physics and Storytelling

The relevance of quantum physics to storytelling is that there is a constitutive role of the storytelling apparatus and measurement processes that needs to be investigated. Karen Barad's (2007) treatise on quantum physics attempts to realign "materiality" and "discourse." Barad (2007, p. 229) turns to Foucault (1977, p. 137) for a theory of discourse, where regimes of power partition "as closely as possible time, space, and movement." Quantum physics calls narrative representationalism and Newtonian interactivity physics into question. It does this by refusing to separate measurement agencies from the constitution of phenomenon being observed in the present moment of being.

Agential Realist Ontology This story brings us closer to what is called the "agential realism" approach to relational process ontology. Karen Barad (2007) looks at the intra-play of materiality and discourse. They intra-penetrate each other. As a physical therapist, she is in touch with the corporeal real body of the patients and she uses her knowledge assemblage of Being. But, what is agential realism and how does it relate to ontological storytelling? Next, we answer to this question.

Agential realism is beyond the boundaries of human and nonhuman or social and nonsocial. "The framework of agential realism does not limit its reassessment of the matter of bodies to the realm of the human (or to the body's surface) or to the domain of the social" (Barad 2007, p. 209). Agential realist ontology views "matter as a process of materialization beyond human and social, into a more complex understanding of normalizing practices (including regulatory ones) and the participatory role in the production of bodies" (p. 210). Barad (2007, p. 393) develops the theory of "agential realism" by reworking the notions of causality and agency. Instead of causal relationships between distinct sequential events, as in narrative, causality is "rethought in terms of intra-activity."

According to agential realism, there are not separate kinds of materiality and so the linkage between discursive practices and their materiality effects on bodies is not all that mysterious: discursive practices are material efficacious, to the extent that they are because there is a *causal* linkage between them, which is to be understood in terms of the causality of intra-actions (Barad 2007, p. 211).

Agential realism looks at the in-betweenness of discourse and materiality in a timespacemattering approach. Betweenness is not about the traditional physics of interaction, but rather the term "intra-action" including the dash (-) is preferred. In the dash, there is an intra-play, even an intra-penetration as discourse and materiality co-mingle. A key concept is "intra-action," which is defined as "mutual constitution of objects and agencies of observation within phenomena" (Barad 2007, p. 197). Intra-action is different from the old physics word, "interaction," which "assumes the prior existence of distinct entities" (p. 197). Neils Bohr, for example, asserted that "*the nature of the observed phenomenon changes with corresponding changes in the apparatus*" (Barad 2007, p. 106).

In this Posthumanist ontology, storytelling is a material phenomenon accomplished by material means and the apparatus of observing and measuring in ways that have a differential mattering in the material world that is more or less agential. Stories are not ideational representations, not narratives, but are occurring in mattering practices. Stories are not abstract objects, not separable from the measuring and productive agencies of storytelling. There is no storytelling apart from the observing and measuring apparatus. It is this entanglement that is the basis of the agential realist ontology of storytelling. Finally, "it is the reiterative character of performativity that opens up the possibility of agency." (p. 213). We are in agreement with the framework and aware of a fair amount of other literature that moves in this direction. Among the works that we have not yet cited and that seem clearly to move in the direction of materiality-storytelling are John Law's *After Method* (Law 2004), Evelyn Fox Keller's *Reflections on Gender in Science* (Fox Keller 1985) and *A Feeling for the Organism: The Life and Work of Barbara McClintock* (Fox Keller 1983), and Don Ihde's *Technology and the Life World* (Ihde 1990).

However, the framework has been subjected to controversy in another body of literature. The works of K. G. Gergen (Gergen and Kaye 1992), J. S. Bruner (1990), and Donald Polkinghorne (1988), which are more focused on social construction and meaning rather than the materiality of story, are noteworthy cases of resistance to the framework.

Power does not determine subject formation in storytelling since, as Foucault observes, there is always conflict and struggle in local acts of resistance (p. 213). In the Foucauldian microphysics of storytelling, we can examine the causality and agency from Barad's agential realism that is a very different paradigm than the usual social construction or interpretivism approach to narrative representationalism. The difference is the focus on the productive and materializing effects of the microphysics of storytelling. This requires a rethinking of causality. Instead of causality as a linear movement from a chain of causes to effect, causality has agential materiality, in conditions that are embodied and contingent, including the agencies of observation, articulation, tonality, and listening in storytelling acts. Every storytelling involves a choice of apparatus that gives attention to some aspects of the phenomenon being included while others are being excluded. We want to pay closer attention in qualitative methods to the conditions and apparatus of observation in accomplishing storytelling.

It is time to challenge the linearity thesis of most narrative work because as Barad (2007, p. 394) puts it, "Future moments don't follow present ones like beads on a string." While narrative is mostly retrospective sensemaking, and rooted in a social construction paradigm that does not address the material world, for Barad (2007, p. 396) "mattering is an integral part of the world in its dynamic presencing."

Presencing and mattering are ethical challenges to traditional narratology because each moment of Being-ness, such as in the explorations of Taekwondo and Physical Therapy, storytelling is alive to the possibilities of becoming (p. 396).

Next, we look more in depth at Latour's Actor-Network-Theory.

7.4.1 Actor-Network-Theory and Living Materiality-Storytelling

Bruno Latour's (1999, 2005) ANT attempts to reverse the ills of social constructionism. To say that something is socially constructed has come to mean that "something was not true …: either something was real and not constructed, or it was constructed and artificial, contrived and invented, made up and false" (p. 90). Latour's preference is to focus on the material ways humans and nonhuman actants are constructed assemblages. The problem with social construction is that it reduces all constructions (or all assemblages) to just one type of material, the social. It is a strange storytelling where only the social does assemblage, and nonhuman materiality has no agency at all. Latour (2005, pp. 52–53) defines agency as "*doing* something, "making some difference to a state of affairs" or "transforming some As into Bs through trials with Cs" (p. 53). Storytelling is critical to agency, since "an invisible agency that makes no difference, produces no transformation, leaves no traces, and enters no account is *not* an agency. Period" (p. 53). The cited works of Law and Ihde are also relevant here.

Like Barad (2007), Latour (2005) does not trust the word "interaction." For Latour, the word, interaction, implies that they are of the same time, in the same place, of the same agency, use the same optic, and have the same pressure. In ANT storytelling, there is an exploration of how materiality is assembled from different times, places, agencies, optics, and pressures.

An example given by Latour (2005, pp. 200–202) is that classroom interaction is not synchronic, not isotopic, not homogeneous, not synoptic, and not isobaric. The material we study is not synchronic, not of the same time, since quite a variety is coming from quotes from books of many times. Materials are not isotopic, in that entities ranging from bodies, desks, windows, flooring, and ceiling to books, clothes, and cell phones come from many different places. Agencies are not homogeneous, such as when notes are made about books, typed with a keyboard into data storage, configured onto a slide presentation, transferred digitally to a projector, and bounced from a screen onto the mind of a half-asleep student. Those in the classroom are not synoptic, not using the same optic or logic. The people are not isobaric, since students have many different pressures, and their pressures are not the same as those of instructors or teaching assistants. Dreyfus challenges the image of the classroom as an organization. He believes that learning communities or communities of inquiry are better images of organizations that break a number of linear structures into other components. In any case, issues of heterogeneous agencies maintain.

In assemblage-materiality-storytelling, we might just discover the "heaps of paraphernalia" (Latour 2005, p. 75) societies and organizations organize in action. As an exercise, take a moment and do an inventory of the assemblage you have assembled around you this day, on your person, and around your person.

7.4.2 Assemblage-Materiality-Onto Story

Material objects and material world has been thought to contain energy for long. However, "socio"/material agency is much more believable nowadays, as we encounter material assemblages operating partly or even sometimes entirely without human agents' interference. Machine to machine (M2M) communication is among the most conspicuous scenes of a material assemblage we usually see without noticing its nonhuman agency. Technology has brought up ways machines can connect to one another, communicate, transfer data and signals, and, on the whole, operate—at least partly—on their own. These M2M communications providing us with a bunch of products—such as the internet, telecommunication, and networks—represent a noticeable exemplar of what a material assemblage as a nonhuman agency produces through its combinative energy.

We connect to the internet via our computers or smart phones, upload and download data, and contact other people whom themselves have connected to the internet via their own devices. We do all these without even noticing how various material assemblages communicate and work together to bring us these possibilities. Just like organizations comprising human agencies collaborate in a supply chain, an industry, or even in broader environments to produce products and services, material assemblages comprising nonhuman agencies collaborate to produce services for humans and human agencies. Organizations using the intranet are instances of human agencies that rely on services partly provided by nonhuman agencies.

In this section, we present an ontological story of M2M networks as a material assemblage to improve understanding of this nonhuman. Firstly, we define the concept of "material assemblage," and then proceed to analyze a M2M network as an agentic assemblage.

Any specific structuration of events and space form an assemblage. As said by Deleuze and Guattari, and Bennett, accordingly, "assemblage is an ad hoc grouping of vibrant materials" with their enhancing and confounding energies. This diverse range of energies leads to its "emergent property" which is not the simple sum of all members' energies. Hence, an assemblage holds a distinct efficacy called "agency of assemblage" (Bennett 2010, pp. 23–24). It is noteworthy to mention that "efficacy is the capacity of things to correspondingly affect and being affected by other things" (Bennett 2010, p. 21). Grouping of different things in an assemblage creates a unique efficacy to the assemblage, which is different from the efficacies of individual parts. This issue is brought up by Spinoza under the title "effective bodies" (Bennett 2010, p. 23).Spinoza believes that all bodies—the term he uses for things—are derived from a common substance. He also asserts that more complex bodies—what we call assemblages—are diverse compositions or forms of simple bodies. Bodies increase their power in a heterogeneous assemblage because the assemblage can affiliate with more types of bodies than a simple body does. Hence, it can affect and be affected with greater bodies, that is, has superior efficacy. From agency point of view, it means that efficacy of things is ontologically enhanced and distributed across a heterogeneous field (Bennett 2010, p. 23). While ontological is widely explored in systems theory a la von Bertalanffy and Gregory Bateson, our focus here is on the intra-activity of discourse with materiality. Plus materiality here is in the sense of quantum physics (Barad 2007). Intra-activity, instead of interactivity, distinguishes our argument from that of system theory.

Most of the things around us are in fact complex bodies that also interact with a great number of other bodies. In other words, agents hardly act alone. Rather, their agency is produced in collaboration and interaction with other surrounding bodies.

M2M is an assemblage of things (here, machines) that influence being affected by other things. The more partner and more diversity it involves, there is greater power affecting other things (or forming a more dominant mode in Spinoza's term), hence approving Bennett's point in enhancing power through heterogeneous assemblage. The result is to make something happen which cannot be performed through each of the machines/things separately. In the M2M case, networks and connections are the emergent property of things assemblage.

Although once created by human agents, M2M assemblages in all forms now work partially or completely independent of human agencies and create desirable or unwanted, planned or unplanned, and even unpredictable consequences. The production capacity of these entirely material assemblages represents energies covert in material objects.

To further explore how a material assemblage entirely consisting of material objects operates as an agency, we analyze a specific type of M2M communication with which we are all familiar, that is mobile communications. In our exploration, we take advantage of an ordinary organization consisting of human actants as a metaphor to compare a material assemblage of nonhuman agencies versus a typical assemblage of human agencies. This comparison help us better appreciate energy and agency of material objects.

When we make a call to another person via our cell phones, we dial a number associated with that person and wait for the result, which might be an answer or a reject from the other party, connecting to his or her voicemail, or even dropping the call.

Are dialing or pressing the "answer" button the only processes and activities required to make a phone call? Obviously not. These are activities we perform, but there are a whole bunch of other procedures and actions that occur as we make a simple phone call, entirely by a material assemblage called the "Mobile Telecommunication Network." Here, we just explore one specific technology using SIM Cards to establish mobile telecommunication.

We observe two machines—our cell phones—are connected and if we observe more carefully, we may notice Base Transciver Station (BTS) towers. But there are still many other machines that communicate to make this connection possible, which we do not see. The below figure provides a simple schematic illustration of the assemblage (Fig. 7.1).

To start a cell phone communication, the subject (here a human agent) has to obtain a cell phone, that is, the human–nonhuman interface of the assemblage. When a subscriber registers his or her equipment on the Global System for Mobile (GSM) network, its data is saved in Home Location Register (HLR). User Equipment (UE) in our example consists of cell phone and SIM Card. Like the organization front line, UE is the part visible to the customers.

HLR is a central database containing all the cell phone information of subscribers who are authorized to use the GSM network. The GSM network is the most prevalent standard for mobile communication worldwide.

As a subscriber attempts to make a phone call, he or she sends the destination phone number to the BTS via a wireless connection. BTS is equipment which prepares wireless communication between UE and the network. The nearest BTS to the cell phone receives the information and pass it to the MSC.

MSC searches for called party information in its own VLR. VLR is a part which determines where other mobile subscribers are located. If VLR does not have subscriber information, a request letter is sent to HLR by MSC to query the called party information. This part of process is comparable to decision making in an organization. If the decision

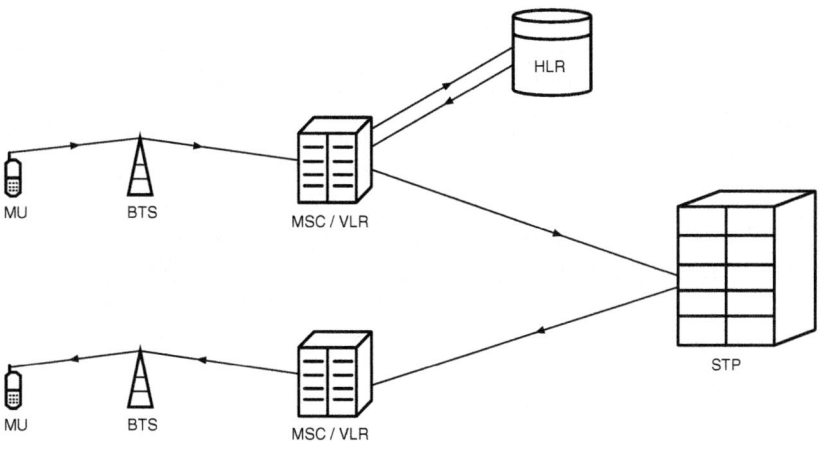

Fig. 7.1 Mobile telecommunication network (Source: Nazanin Tourani original)

regards a recurring issue, the decision maker may refer to the related records and find out the results. However, when new circumstances arise, the same person may refer to higher managers in hierarchy to ask for additional information and guideline.

When HLR returns the requested information to MSC, first MSC sends a message to the destination MSC, related to the place where the called party is located. This message is sent through STP, which is responsible for transferring signals between GSM network nodes. The destination MSC receives and sends the message to the destination BTS. Finally, BTS sends a message to the called party and the called party's cell phone rings. At this point, a signaling circuit establishes between the called party's UE and the calling party's UE via MSC and STP. When the called party presses the answer button, a voice channel will be reserved between these two subscribers to make a voice call. After establishing the voice channel, the signaling circuit will be released to be used for another voice call.

The call flow here is easily comparable to the data flow and communication channels in an organization. HLR, MSC, BTS, and UE represent the hierarchy of the assemblage. It is interesting that in this material assemblage, like most organizations, the more information a part has, the higher position it holds.

In our final section, we turn to indigenous peoples' storytelling and a more-living materiality to praise Native American storytelling, which has a very special relationship to the materiality. The implications for drawing together agential realism and our path, with the "native American" storytelling is that we can look to methodologies where embodiment is more than how material conditions are represented. We can go to looking at the intra-play of storytelling with a living materiality.

7.5   Indigenous People's Storytelling and Living Sociomateriality

Living Sociomateriality Indigenous storytelling is about a live-centered participating with a living materiality world. Cajete (2000, p. 2) believes that a new science is needed, a "native science" which he defines as "coming to know" that tribal processes "have evolved through human experience with the natural world. Native science is born of a lived and storied participation with the natural landscape." "Native science" for Cajete (2000, p. 14) "is not quantum physics or environmental science, but it has come to similar understandings about the workings of the natural laws through experience and participation with the natural world."

A special sense of materiality, what Vizenor (1998, p. 15) calls "*transmotion,*" is defined as "that sense of native motion and an active presence [that] is sui generis sovereignty" and "a reciprocal use of nature, not a monotheistic, territorial sovereignty." "The transmotion of ledger art is a creative connection to the motion of horses depicted in winter counts and heraldic hide paintings." (p. 179). Storied transmotion is a material "presence in stories, an actual presence in the memories of others, and an obviative presence as semantic evidence" and in a Bakhtinian sense "a dialogical circle" (p. 169) and "in a 'dialogical context,' the conversions of [ethical] answerability" (p. 27, bracketed addition, mine). Stories are a transmotion and a virtual sense of presence in animated and embodied native memories that includes in a Posthumanist philosophy, animal memories (p. 170).

Cajete (2000, p. 21) wants to reclaim an indigenous view of "animism" from its dismissal by modernism. "Native science is an echo of a pre-modern affinity for participation with a non-human world." (p. 23). Not just an affinity for animal participation, but an entire basophilic instinct of living with all other living things. "Life and death are transformations of energy into new forms, the material and energetic fuel of nature's creativity." (p. 21). This sentiment echoes Barad's Posthumanist energetic view of quantum physics in her respect for the proton lifeworld in which all things are actively participating with all other things. In the "animistic logic of Indigenous, oral peoples," objects such as stones and mountains "are thought to be alive and from whom certain names, spoken out loud, may be felt to influence the things or beings that they name, for whom particular plants, particular animals, particular places, person and power may all be felt to 'participate' in one another's existence" (Levy-Bruhl 1985, cited in Cajete 2000, p. 27). This work by "native American Indian" writers and scholars speaks to an intra-active relationship between storytelling and materiality of living ecological places. Here, we encounter a different sort of storytelling, a storytelling of living in an enlivened material world. Works by Cajete, Vizenor, Sarris, Allen, Silko, Womack, and Warrior challenge the Western

European-American approaches to lifeless narrative representationalism by noting our participation in wider orders of nature.

Next, we look at the Native American storytelling scholars who we believe are saying something important about the relationship of storytelling and materiality.

A growing number of Native American scholars have continued to break away from the dictates of western narratives (e.g., Allen 1992; Cajete 2000; Owens 2001; Silko 1981; Vizenor 1998, to name but a few). And not just Native American, but also indigenous writers around the world are challenging western narrative hegemony. Smith, from New Zealand, for example, rejects "the idea that the story of history can be told in one coherent narrative" (Smith 1999, p. 31). To decolonialize for Smith (1999, pp. 34–35) means recovering stories of the past that were marginalized, "telling our stories form the past, reclaiming the past, giving testimony to the injustices of the past" as strategies "employed by indigenous peoples struggling for justice."

Our sense of time, space, and matter are largely determined by society. From this socialization, expectations of how our storytelling is to be constructed are situated. Our emphasis is on the materiality focus on storytelling (transmotion, survivance, Earth-based) which has been marginalized in Western narratology. In this regard, we refer to nonlinear storytelling examples, such as Joseph Heller's *Catch-22* (Heller 1961), Irvine Welsh's *Trainspotting* (Welsh 1993), and James Joyce's *Ulysses* (Joyce 1922), and *Finnegans Wake* (Joyce 1939) among many others.

Time Paul Ricoeur (1984, p. 62) asserts that "within-time-ness or being – 'within' – time deploys features irreducible to the representation of linear time." Among Native American scholars, Highwater (1981, p. 92) points to an errant western obsession with materiality of time: "There are a great many preconceptions with which various societies grasp the world and make it comprehensible, but in the West there is perhaps no single idea as obsessive as the notion of the material reality of time." When we grasp the materiality conception of time in the West, we begin to grasp the difficulty storytellers and their audiences have in moving away from mechanical clock and associated linear succession conceptions of time. "To designate clock-time as 'real time' is the same as calling

American money 'real money.' ... It is the imperative in the West to fal-sify our consciousness so it fits the stream of duration that carries us out of the past and into the future, the modality of linearity, and it is com-posed equally of a past, present, and future through which a sequence of enduring events follow one another in an orderly and calculable manner" (Highwater 1981, p. 94).

Western concepts of time, space, and matter have created a gap between linear narrative and native peoples' more nonlinear storytelling. Indigenous peoples' scholarship has long suspected that western narrative is trying to keep people from noticing that the Posthuman world is decid-edly nonlinear. It takes a good deal of special schooling, and media pro-paganda, to convince the masses that they live in flatland. Most narrative is about past remembrances strung together in a linear sequence of events. These remembrances are folded onto the future, where some expect the past to replicate its flat lines onto the future. Indeed, there are many people and contexts in which we don't think of time as linear. Nonetheless, it is the dominant, educated, cosmopolitan model.

When released from clock-time, we find our temporal experience much more personal and relativistically constructed. We are better able to focus on now, as an extended moment, where time appears to slow and speed up. In those moments, we can glimpse, for a fraction of a second, that life is not running along the presumed linear trails.

Space Bergson noted that our Western conceptions of time are fixated on space "If we want to reflect on time, it is space that responds." Thus, "duration is always expressed as extension." And "the past is understood as something lying (physically] *behind* us, the future as lying somewhere *ahead* of us" (As cited in Highwater 1981, p. 96). Leslie Silko (1981) writes out of her Laguna Pueblo tradition of storytelling in which according to Highwater (1981, p. 100) "there is always an intimation of a reality larger and essentially different from that of the dominant soci-ety and its predominantly naturalistic literature." Cajete (2000, pp. 210–211) describes indigenous causational paradigm as a living hologram where one moves around a deeply internalized sense of place filled with life, looking from different animal and plant perspectives together with looking at landscape from the north, south, east, west, below, above, or from within, in order to learn to bring the different understandings of living community together. Note how different this living hologram of ecological consciousness is in comparison to Western Newtonian-Cartesian –mechanistic science and capitalism paradigms of "a mass of 'dead' matter ripe for manipulation and material gain" (Cajete 2000, p. 212).

We would like to close our chapter on the qualitative arts of storytelling-materiality with David Boje's account of becoming a blacksmith. It is also a summary of onto-storytelling as a method.

Example 7.5 David Boje's Account of Becoming a Blacksmith

I am a blacksmith and a storyteller. I prefer a storytelling that is ontological, about real stuff, metal things, coal, fire, water, and tools. I learned about onto-storytelling from Jane Bennett (2010), who applies Bruno Latour (1999, 2005) Actor-Network-Theory (ANT). An onto-storytelling is about humans in relation to material things, each of which is an actant in the storytelling. In ANT, there are material actants, human agents, and they have more or less agency in a situation. The onto-storytelling is about how humans, animals, and material things are in intra-activity. We use the word "intra-action" instead of the word, "interaction" because we want to invoke the new quantum physics, where things and people do not interact, but at a molecular-proton level, materiality crosses in-between. Karen Barad's (2007) work, for example, looks at the intra-activity of materiality and discourse. Storytelling is a domain of discourse. So here, in this chapter, we want to see in onto-storytelling, how the storytelling and the materiality cross-between, intra-twine, and intra-play. We will tell you more about quantum physics, and the new ways of looking at materiality, as we proceed with our examples of onto-storytelling.

As I was telling, I have become a blacksmith. More accurately, I am in the process of becoming one. I live in Las Cruces, New Mexico, on a small horse ranch. I am the first one in my family tree to attend college, much less get a Ph.D., or become a professor, and do all sorts of theory things. I have been interested in storytelling for 30 years, but it's only since I took up blacksmithing that I became fascinated with the ontological aspects.

About a decade ago, while she was still living, my mother Loraine Joyce told me about my great grandfather and how he had become a blacksmith and livery stable owner in Goldendale, Washington. I suppose that planted a seed to become a blacksmith. She also told me that on my mother's and dad's sides of the family, someone had married Native Americans. It was the first I had ever heard of it.

My mother took down a shoebox full of photos from a closet shelf. She was cursing as I began tearing through the stack. "These photos are not worth anything. I hate this part of my life. I want to forget about it." "What is my grandmother's name," I asked. I always knew her as grandmother. "Her name was Wilda," my Mother screeched with more emotion than I was used to seeing. She then tore a photo of my grandmother Wilda. I pieced it back together and took a digital photo. You can see the diagonal tear through Wilda's hat and through her horse.

My great grandfather, on my mother's side, is William Henry Shelton (born Jul 26, 1863, in Brownstown, Indiana; died Aug 18, 1946, in Toppenish, WN, buried at Tacoma Cemetery in Yakima, WN). William Shelton became a blacksmith. He and his wife Virginia Tuttle (born Mar 13, 1863, in Wayne County, Iowa, and died Oct 11, 1944, in Toppenish and buried in Yakima, WN) came to Washington in a covered wagon on the Oregon Trail in 1897. One of their children (Henry Wayne) died on that trail. The largest migration along the Oregon Trail took place in the 1840s with travel continuing until the transcontinental railroad was completed in 1869. William and Virginia Tuttle could not afford the train, so they did the 2,000 mile trek by wagon in 1897.

William and Virginia were wheat farmers, but when they could not make a go of that, they moved to the town of Goldendale and opened that town's first livery stable and blacksmith shop around 1900. My grandmother, Wilda, was born on Jan 28, 1902. She had two older sisters and two older brothers (counting the one that died on the Oregon Trail), and one younger brother, named Joseph Gerald Shelton (born Jun 13, 1904; died Aug 18, 1937). He went by the name Gerald and married (or may have) Stella LaClaire, a Native American. My mother pulled this photo out of the shoebox, and I made a digital image.

Their children included Georgia (or Georgie) & perhaps a Darlene; Gerald worked in circus and Rodeos; he as an alcoholic was beaten by the police, and died in the Toppenish jail, WN; the story is they threw him in so many times, they got tired and just beat him to death.

About three or four years ago, I decided to become a blacksmith. I was born after my great-grandparents had died. So I can only imagine what their blacksmith shop was like. I drove to Goldendale and poked about, but could not find any records of it.

Most storytelling has degenerated into just retrospective narrative, into epistemology disconnected from the material world. I prefer to do some other storytelling in addition to narrative work. In particular, I like the living stories and the antenarratives. A living story is part of a web of living stories, in the present, in what Mikhail Bakhtin (1993) calls the once-occurrent event-ness of being. It is this living story, caught up in the now, that for me is closest to the ontological. Finally, in antenarrative, a term I invented in a 2001 book, we are dealing with the future. Antenarrative is a double meaning: "ante" as a 'before' narrative becoming fossilized, and "ante" as a "bet" on how transformation is happening.

I use my blacksmithing to explore my own living story, as I am a black-smith. I also explore what I call the "spiral" and "assemblage" antenarra-tives. Most antenarratives that have been studied are linear or cyclic, and so the spiral and assemblage ones are quite novel. In a linear or cyclic antenar-rative, the narrative has its way, just folds the past onto a future that is expected to replicate. Spirals and assemblages happen differently. They are antenarratives that fold the present moment of Being-ness (living story, if you will) onto the future.

As someone becoming a blacksmith, I am in touch with my living story and with materiality, with coal, water, fire, air to make the fire hotter, and the iron I am holding with tongs. You know it's real, a living flesh and blood story, when you forget to be in the now and grab a piece of grey iron that looks cool enough, but is still 1200 degrees. It melts the skin on your fingers like butter. You just stare at your fingers and cannot believe you just did something so stupid. It's so hot, you don't even feel it. You know that the pain is coming. The pain comes in an hour or so when the blisters form and continues for the next few days. I apply lots of Aloe Vera and meditate on the importance of a blacksmith staying in the moment.

Materiality for blacksmiths is different than for non-smiths. To a smithy, metal is only solid when it's cold. Heat it up to 1900 or 2000 degrees, and you can bend it, move it, and forge it into about any shape you want. Just let it cool before you touch it. Before you can shape metal, you need the coal fire to be just the right heat, to form some coke, keep the slag out of the heart of the fire, and keep tending to it so the sides of the coal fire retain the heat. As you heat up a piece of iron, you learn to pay attention, or the iron will overheat, becoming yellow-white and shooting sparkles just before it melts. As you heat the iron above a red heat, you can begin the hammering. That hammering on the anvil actually changes the carbon structure, moving the molecules into a tighter configuration, strengthening the metal. Each type of metal has different heating and bending properties. And it takes quite a while to learn to tend the coal fire, watch your piece, learn to hold it steady with a pair of tongs, and understand how to use a hammer. Using a hammer sounds simple, but when you are holding hot iron with tongs, and wanting to shape a leaf, you learn it's far from a simple thing.

I have been practicing making leaves on the ends of ½ inch and some-times 5/8th round stock. I bend the stalks of the leaves into something life-like, and one of them I leave longer, so it spirals around the others. I have to make at least 100 leaves before they will be anywhere near the quality I want to achieve. To get good at something, you have to do it again and again, until you master it.

My blacksmith shop is fashioned out of straw bale, covered in chicken wire, and coated with stucco. It's unlikely that William Shelton used straw bale construction, but it's a possibility. My shop has a metal roof and two eight-foot doors that swing easily on strap hinges. I kept most of the floor earthen since it is easier on the feet than standing on concrete. And you don't want to drop red or white hot metal on concrete. To be a blacksmith, you need all sorts of tools. Some I bought: hammers, tongs, and a post vice. Others I borrowed, such as a fine anvil, from my neighbor, Pep Gomez. Pep is a master blacksmith and teaches the metal arts at Dona Ana Community College. He helped me fashion the metal strap hinges holding the doors. I finished them and mounted them, when my son Raymond was in town. Other tools I made: a coal forge, some more tongs. You can never have enough tongs. I also made some rather large power hammers out of recycled truck axles, leaf springs, and some huge I-beams I got from a salvage yard.

It took me a year before I could make a leaf that looked anything like a leaf. I also make knives out of railroad pikes and horseshoes. I cold-forged wire into spiral pens, smaller knives, swords, and hearts. I fashioned a sword for my son Raymond out of leaf spring. I sold one of the railroad spike knives for $75 at an Arts Convention in Las Cruces in 2009. The others I gave as gifts to relatives and my students.

Last November, I went to Gunter's weeklong blacksmith workshop in Moriarty, New Mexico. Besides how to properly tend a coal fire and various ways to twist metal, eight of us learned to forge weld. Forge welding is when you heat pieces of metal to a white-yellow color and try to get them to stick together. It takes perfect heat, perfect timing, and just the right amount of hammer blow to make a forge weld. It helps to use some borax, applied at a red heat, to keep the pieces clean before you try the forge weld.

References

Allen, P. G. (1992). *The Sacred Hoop: Recovering the Feminine in American Indian Traditions*. Boston: Beacon.

Bakhtin, M. M. (1993). *Toward a Philosophy of the Act* (M. H. Liapunov, Ed., & V. Liapunov, Trans.). Austin: University of Texas Press.

Barad, K. (2007). *Meeting the Universe Halfway: Quantum Physics and the Entanglement of Matter and Meaning*. Durham/London: Duke University Press.

Benjamin, W. (1963). The Story-Teller: Reflections on the Works of Nicolai Leskov. *Chicago Review, 16*(1), 80–101.

Bennett, J. (2010). *Vibrant Matter: A Political Ecology of Things*. Durham: Duke University Press.

Bruner, J. S. (1990). *Act of Meaning*. Cambridge, MA: Harvard University Press.

Boje, D. M. (2008). *Storytelling Organizations*. London: Sage.

Cajete, G. (2000). *Native Science; Natural Laws of Interdependence.* Santa Fe: Clear Light Publishers.

Campbell, M., & Gregor, F. (2004). *Mapping Social Relations: A Primer in Doing Institutional Ethnography.* Walnut Creek: AltaMira Press.

Dreyfus, H. L., & Dreyfus, S. E. (1986). *Mind Over Machine: The Power of Human Intuition and Expertise in the Era of the Computer.* Oxford: Basil Blackwell.

Fairhurst, G. T., & Putnam, L. (2004). Organizations as Discursive Constructions. *Communication Theory, 14*(1), 5–26.

Foucault, M. (1977). *Discipline and Punish* (A. Sheridan, Trans., p. 242). New York: Vintage, 1979.

Fox Keller, E. (1983). *A Feeling for the Organism: The Life and Work of Barbara McClintock.* New York: Freeman.

Fox Keller, E. (1985). *Reflections on Gender in Science.* New Haven: Yale University Press.

Gergen, K. J., & Kaye, J. (1992). Beyond Narrative in the Negotiation of Therapeutic Meaning. In S. McNamee & K. J. Gergen (Eds.), *Therapy as Social Construction.* Newbury Park: Sage.

Heller, J. (1961). *Catch-22.* Simon & Schuster.

Highwater, J. (1981). *The Sun, He Dies.* New York: Lippincott & Crowell.

Holstein, J., & Gubrium, J. (1999). *The Self We Live by: Narrative Identity in a Post-modern World.* Oxford: Oxford University Press.

Ihde, D. (1990). *Technology and the Lifeworld: From Garden to Earth.* Bloomington: Indiana University Press.

Joyce, J. (1922). ULYSSES. Dijon: Sylvia Beach.

Joyce, J. (1939). *Finnegans Wake.* London: Faber and Faber.

Latour, B. (1999). *Pandora's Hope: Essays on the Reality of Science Studies.* Cambridge: Harvard University Press.

Latour, B. (2005). *Reassembling the Social: An Introduction to Actor-Network Theory.* New York: Oxford University Press.

Latour, B. (2007). *Reassembling the Social.* Hampshire: Oxford University Press.

Law, J. (2004). *After Method: Mess in Social Science Research (International Library of Sociology).* New York: Routledge.

Myers, M. H. (2009). *Institutional Ethnography: How Tenured Academic Women Discuss Success.* UNM dissertation Open Source available at http://repository.unm.edu/bitstream/handle/1928/10353/MHM%20Dissertation%20Institutional%20Ethnography%202009.pdf?sequence=1

Owens, L. (2001). As if an Indian Were Really an Indian: Unamericans, Euaamericans. *Paradoxa, 15*, 170–183.

Polkinghorne, D. E. (1988). *Narrative Knowing and the Human Science*. Albany: State University of New York.

Ricoeur, P. (1984). *Time and Narrative* (K. McLaughlin & D. Pellauer, Trans., pp. 1, 64). Chicago: The University of Chicago Press.

Silko, L. M. (1981). Language and Literature from a Pueblo Perspective. In *English Literature: Opening Up the Canon* (pp. 54–72). Baltimore: Johns Hopkins University Press.

Smith, D. E. (1990, 2002/2003). *Texts, Facts, and Femininity: Exploring the Relations of Ruling*. London: Routledge, 2007.

Smith, D. E. (1999). *Writing the Social: Critique, Theory, and Investigations*. Toronto: University of Toronto Press.

Smith, D. E. (2005). *Institutional Ethnography: Sociology for People*. Boston: Northeastern University Press.

Vizenor, G. (1998). *Fugitive Poses: Native American Indian Scenes of Absence and Presence*. Lincoln: University of Nebraska Press.

Welsh, I. (1993). *Trainspotting*. London: Secker & Warburg.

8

Interpretation, Reflexivity and Imagination in Qualitative Research

Yiannis Gabriel

8.1 Introduction

Quantitative and qualitative researchers in the social sciences face fundamentally different challenges in the ways they seek to generate and test new knowledge. Quantitative researchers aim for law-like generalizations after the model of the natural sciences. They seek to establish links between causes and effects that apply irrespective of time and place and come up with accurate generalizations. These can subsequently be used as the basis for making predictions. "A rise in unemployment levels causes a rise in crime" may lack the certainty and precision of "A rise of the temperature of water to 100 degrees Celsius causes it to boil, changing from liquid to gas", but the two are fundamentally similar types of statements. Both of these statements may be qualified. "Water boils at 100 degrees at sea level" or "Crime level rises as a result of unemployment, unless a government invests in youth training schemes". Both of these statements

Y. Gabriel (✉)
Bath University, Bath, UK

© The Author(s) 2018
M. Ciesielska, D. Jemielniak (eds.), *Qualitative Methodologies in Organization Studies*,
https://doi.org/10.1007/978-3-319-65217-7_8

may be falsified through evidence. Ultimately, however, they both stand or fall depending on whether the relation between different variables can be established with a degree of rigour demanded by the methodology of the disciplines in question.

Qualitative researchers confront a different type of challenge. The statement "Faced with job loss and financial ruin, Accountant X turned to fraud, blackmail and eventually committed suicide" aims at explaining the experiences and actions of a specific individual faced with a specific predicament. Unlike the quantitative researcher who seeks to understand the particular as an instance of the general, the qualitative researcher is seeking to discover the meaning of particular events and experiences, aiming to understand these phenomena as outcomes, intended or unintended, of *meaningful* human actions, emotions and intentions.

Both quantitative and qualitative researchers seek to answer the question 'why', but they approach it from different angles. Quantitative researchers approach it from the angle of causality, while qualitative researchers approach it from the angle of purpose and intent. In the social sciences, the two approaches may address what appear as the same social phenomena, like suicide, poverty or illness, and the same organizational phenomena, like leadership, conflict or change, but they do so with different agendas, interests and logics. Quantitative research is guided by the logic of explanation, verification and prediction even when its explanations prove temporary, provisional or even unattainable. Qualitative research, on the other hand, is guided by the logic of understanding by constructing plausible, meaningful and coherent narratives that account for the experiences and the actions of different individuals and groups. Quantitative research looks at human actions as behaviour, not so very different from the behaviour of animals or natural objects; qualitative research, on the other hand, looks at human actions as purposive, meaningful and emotional.

The distinction between logico-scientific knowledge, which is the aim of much of quantitative research, and narrative knowledge, which is the aim of qualitative research, was first put forward by Bruner (1986) but may be traced to Habermas (1972) and others who argue that they each subscribe to different ways of establishing the truthfulness of their claims. Logico-scientific knowledge, which is sovereign in the natural sciences, aims for rigour and generalizability. Narrative knowledge, on the other hand, makes no claim to generalizability and its predictions are highly contingent and qualified. "The king died, then the queen died of grief" does not purport to explain what happens to all queens following the death of their husbands, but may offer a very sound narrative explanation for the death of a particular queen who died shortly after her beloved husband. Likewise, "the pilots were involved in a heated conversation about the imminent merger of their company and over-ran the runway, landing the plane on an adjacent field" may not amount to a general explanation of missed landings or pilot conversations but may explain the particular incident at hand. What is now known as 'narratology' does not seek explanations in fixed relations between causes and effects but in establishing the links between people's actions, purposive and accidental, and their outcomes by placing them in plausible plots.

Narrative knowledge and logico-scientific knowledge call for different ways of thinking and different ways of establishing the truthfulness of their claims, but they are not unable to interact with each other, challenge each other and, at times, support each other. A qualitative ethnographic case study, for example, may clarify concepts and suggest hypotheses that may subsequently be tested through quantitative research (Eisenhardt 2002). Quantitative research, for its part, may generate insights for aggregates of populations that may then be qualified and refined though qualitative research into specific types of cases and phenomena. Thus the statistical argument that job loss is associated with family break-ups, alcoholism and depression may be qualified for instances of people who after losing their job discovered ways of fashioning better lives for themselves and their families. The same statistical relation, however, may account for the fact that Finance Officer X, having lost her job, was more liable to suffer family break-up, alcoholism, depression and suicide.

This chapter examines three fundamental practices of qualitative researchers. These practices call for specific skills and the competence with which qualitative researchers accomplish them determines to a substantial extent the quality of their research. These practices are *interpretation,* which involves a skill at risk of atrophy under the influence of coding software; *reflexivity,* which has become enormously fashionable and risks becoming a box-ticking ritual; and *imagination,* without which no qualitative research has much value or interest but which is at risk of extinction as academic publishing becomes increasingly formulaic and sterile. The chapter deals with how different interpretations may be generated and tested or corroborated. It then examines the role of reflexivity as an "interpretation of interpretation" (Alvesson and Sköldberg 2009) and the extent to which reflexivity can corroborate the validity and enhance the value of an interpretation. We conclude by arguing that, as important as reflexivity is, it should not be treated as the gold standard guaranteeing the quality of qualitative research. Instead, we highlight the role of creative imagination in generating and testing empirically-based insights in qualitative research.

8.2   Interpretation and Hermeneutics

Interpretation, the quest for the meaning of a phenomenon, has long been seen as the heart of most qualitative research. Hermeneutics is the long tradition that examines how to uncover the meanings, often hidden and covert, of different *texts*. Its etymology derives from the Greek word ερμηνεία (interpretation), itself related to the god Hermes, known ambiguously for his ability to convey messages but also to extricate himself from difficult situations (hence, the related word 'hermetic') and to deceive or trick others. Hermeneutics emerged originally as the art of deciphering the deeper meanings of sacred texts, like the Bible, but eventually came to be seen as the quest for deeper meanings in all 'texts' and subsequently human actions, practices, utterances, artefacts and institutions.

Interpretation is a vital part of all qualitative research, as researchers engage with the meaning of narrative, interview materials, images and artefacts and indeed any action or experience whether at individual or collective levels. Interpretation is often a difficult process requiring skill and experience. Even a simple word like "No" can have different meanings depending on the circumstances and the way in which it is spoken. For instance, a "No" may indicate "No, don't talk about this painful subject" or "No, I don't want to talk to you" or "Oh my god, this is incredible" or even, in some circumstances, "Yes, this is exactly what happened, but I didn't expect you to realize". Thus, whether interpreting a particular advertisement, a major political event, a cultural institution or a single individual's actions or experiences, an interpretation is pieced together by identifying different clues, details, signs or symptoms and trying to establish their deeper significance with a new narrative that accounts for all of them. A fundamental principle of interpretation is the *hermeneutic circle* which describes the process of moving from the parts to the whole and back to the parts as a way of building up compelling explanations (Gadamer 1975; Taylor 1971, p. 14). In this way, the interpretation of a story tries to fit different elements like plot, characters, moral message and so forth within a plausible meta-story which leaves no loose ends or unanswered questions. This circular character of interpretation suggests that no interpretation is ever complete or total.

Qualitative researchers are frequently challenged to state the criteria by which they corroborate or support their interpretations. How can judicious and insightful interpretations be distinguished from fanciful, wild or spurious ones? And how truthful can an interpretation be when it is explicitly rejected or denounced by the subject whose actions or experiences it seeks to explain? Consider, for instance, the case of Rose, an anorexic young woman, whose condition becomes life-threatening and her parents agonize about whether to have her hospitalized. Different interpretations may be offered to explain her condition, like:

1. She is a fashion victim, emulating what she sees as ideals of feminine beauty in fashion magazines and the fashion industry;
2. She is seeking to delay the arrival of puberty and remain a child forever;
3. She is rebelling against a society obsessed with food and consumption;
4. She is trying to guilt-trip her parents for constantly criticizing her appearance and/or always privileging her older brother; or, alternatively, she is trying to punish a boy who rejected her;
5. She has succumbed to a death drive and is seeking to kill herself;
6. She suffers from a mental disorder whereby her mind makes her think she is obese, no matter how objectively thin she is;

7. She is starving herself as an excessive form of mental discipline and self-control;
8. She is imitating a particular celebrity with whom she and her friends at school identify.

These and many other interpretations *may* account for Rose's anorexia, but how are we to decide if any of them are valid? It will be immediately clear that the quest for the meaning of an individual's anorexia is a complex one and one that cannot be totally disentangled from the world views and the interest of whoever is proffering an interpretation.

The task of interpretation has been likened to that of a detective who pieces together the evidence in order to reveal the underlying nature of a crime (Ginzburg 1980). It also has similarities with the task of the medical practitioner seeking to diagnose an illness from a patient's different symptoms. In both of these cases, considerable skill is required in identifying different clues and details as meaningful and significant and then moving to the general explanation that informs a search for new clues and details. A successful interpretation is capable of creating an experience of brilliant enlightenment. What was dark and confused suddenly becomes meaningful and clear. What seemed arbitrary and incomprehensible suddenly becomes inevitable and obvious. However, clever or imaginative interpretations are not necessarily true. In some instances, 'wild' interpretations may appear truthful, only to disappoint later. A mother may be told and believe that her son's asthmatic attacks are a response to her husband's sudden explosions of temper but these may later turn out to be caused by a particularly dusty mattress. Wild interpretations can often hit home with vulnerable, sick and gullible people. It is not accidental that some of the most skilful practitioners of interpretation can be found among astrologers, spiritualists, coffee-ground diviners and fortune-tellers, all of whom seek to read large meanings about the past and the future from relatively simple signs.

Interpretation is a difficult art that calls for practice, sensitivity and imagination. Corroborating or supporting particular interpretations is especially hard. This has led many, if not most, qualitative researchers to avoid risking interpretations altogether. The commonest way of doing so is through a practice that has become very common in qualitative papers, namely quoting snippets of interviews or conversations with different respondents in a decontextualized and fragmented way and then tabulating these in a manner that ostensibly indicates dominant themes and patters. This practice is usually supported by the use of qualitative research software, like nVivo, that all but replace interpretation with formulaic techniques of coding text. The art of interpretation is thus in danger of atrophying or even disappearing as qualitative research software enables researchers to organize and code large amounts of qualitative material, seeking to distil from it some theoretical essence. The use of such software in qualitative research has now come to play the part that number-crunching plays in quantitative research. It amounts to a formulaic set of techniques that provide scientific legitimacy for the process whereby large volumes of words are processed to extract something of theoretical value.

Against the interpretive timidity and conservatism represented by coding software, Peter Svensson (2014, p. 174) advocated what he calls 'overinterpretations' which "make it possible to transgress the positive data (i.e. all that which can be heard, observed and perceived on the field) and established ways of representing of social reality". Such overinterpretations seek to reinstate the daring and critical quality of really original interpretations, if necessary by doing violence to conventional meanings and sensemaking attributed to everyday phenomena. Violence, Svensson claims, "persistently tries to see through the most superficial levels of meaning—i.e. what people say and what seems to happen in an organization—in order to get down to the layers of soft and silent power of social life. 'Violence' is a strong but nevertheless an apt term for describing the nature of the overinterpretation." (179) Violent interpretations are liable to arouse hostility and defensive reactions from others including the subjects whose actions and experiences are being interpreted but also other researchers who may dismiss such interpretations as unfounded or wild. What means then are available to qualitative researchers who wish to be daring but also find strong corroboration for their interpretations?

No interpretation can be final or complete. Whether interpreting the meaning of a work of art, a historical event, a biographical detail or text of any sort, new evidence may always emerge which shifts its meaning. Even if no new evidence comes to light, new interpretations may emerge

that supplant or over-ride earlier ones. Yet, this inability to reach defini-
tive interpretations does not make every interpretation equally meaningful
or valid. An interpretation may be original, clever, perceptive, incomplete, misleading or even plain wrong. How then can we test our interpretations and increase our confidence that we have hit the mark? While
no interpretation can be 'proven', every interpretation can be *corroborated*
through a variety of techniques. The internal consistency of an interpretation is obviously paramount, where the interpretation of parts is consistent with the interpretation of the whole. Different signs or clues *all*
pointing in the same direction. In this regard, skilled qualitative researchers act as devil's advocates against their own interpretations, seeking to
identify even a single clue that would undermine or discredit their interpretation. Although not falsifiable on the grounds of individual pieces of
evidence, strong interpretations make clear what evidence would lead to
their refutation. Second, in strong interpretations, specific outcomes are
over-determined. Not only do different signs all point in the same direction, but different mechanisms can be established leading to the same
outcome. Thus, to return to Rose's anorexia, she may indeed be trying to
punish herself for several different reasons at the same time *and* be discov-
ering different ways of punishing herself, including through her anorexia;
for instance, she may have a long history of masochistic and self-harming
tendencies. Third, strong interpretations will generally address, account
for and supersede less strong ones and, even more importantly, will eliminate counter-interpretations. Eliminating all counter-interpretations is
an effective way of corroborating an interpretation, especially if these
counter-interpretations are not 'straw men' but well considered and plausible ones. This is often forgotten by interpretive researchers who sometimes fall in love with their own interpretations and fail to notice that
alternative interpretations of a phenomenon or an event may provide a
stronger, simpler or more complete explanation.

In generating different interpretations, testing, them, qualifying them,
corroborating them and discarding them as qualitative researchers, we are
operating in a somewhat similar manner as natural scientists seeking to
test different hypotheses. Unlike natural scientists, however, our aim is to
generate robust interpretations rather than rigorous and generalizable
chains of cause and effect. The quest for motives and meanings is quite

different from the quest for causes and effects, not least in that the researcher's own motives and meanings can easily become part of the inquiry. More generally, the knowledge generated by the social and human sciences, including theories and concepts, readily become part of the reality under investigation. The object of the human and social sciences is thus a contingent entity in constant change, a change in which ideas and practices resulting from social science itself play a part.

8.3   Reflexivity

If there is no royal road for qualitative research, there is also no full-proof way of validating it. The quest for rigour and objectivity, vital in quantitative research, can hardly be applied to qualitative work. This opens up qualitative research to charges of being ultimately unscientific. It is against this background, that in recent years, there has been an increasing emphasis on reflexivity as an ongoing process of vigilance and self-questioning that qualitative researchers must exercise in order to enhance the trustworthiness and value of their work (Alvesson and Sköldberg 2009; Willig 2001). Reflexivity refers to a constant questioning on the part of qualitative researchers of their own stance vis-à-vis their empirical material, analysis and theorizing. Reflexivity emanates from a realization that the investigators' own values, experiences and motives cannot be separated from the research process. These can be a source of bias and error, which reflexivity may forestall. More importantly, however, the qualitative researchers' own previous experiences, sensitivity and self-knowledge can be valuable resources, enhancing their engagement with their empirical material and deepening their understanding of its meaning and significance. Unlike most quantitative types of research, qualitative research, whether conducted through interviews, observations or even textual analysis prompts the researchers to engage emotionally with their informants, inviting researchers to respond empathetically with the experiences of other people and groups.

Unlike the natural sciences, in the human sciences, the object and the subject under investigation frequently overlap or even coincide (as in the case of autoethnography). This overlap between the knower and the known, the subject who looks in the mirror and the image that stares back from the mirror, has made reflexivity something of a gold standard for qualitative researchers who do not adopt the natural sciences as their model and reject positivist methodologies and traditional criteria of rigour, reliability and validity (e.g. Alvesson and Sköldberg 2009; Cunliffe 2003; Hardy et al. 2001; Hibbert et al. 2006; Tsoukas 1992). A reflexive practice is sometimes used interchangeably with reflective practice, a term popularized by (Schön 1983), as a professional's ability to step back from an immediate activity or experience and turn it into an opportunity for learning. However, as Hibbert et al. (2006) and Malaurent and Avison (2017) have argued, a useful distinction can be made between the two terms. Reflexivity goes beyond mere reflection or reflectiveness in as much as it is 'recursive'—a reflexive practice is one that constantly redefines the practice, through the ability of human statements to alter the state of what is being stated and the person who states it. More generally, a reflexive activity is one in which subject and object co-create each other—in carrying out a piece of research, I create myself as a researcher.

A reflexive researcher recognizes that what she says or writes influences and redefines that about which she is writing as well as herself as the author. This recursive quality of reflexivity is admirably captured in Alvesson and Sköldberg's (2009) ingenious definition of reflexivity as "the interpretation of interpretation" (Alvesson and Sköldberg 2009, p. 9). The implications of this definition are quite far-reaching—reflexivity makes every interpretation potentially unstable since interpretation is liable to change the object being interpreted as well as the subject carrying out the interpretation. Reflexivity, therefore, is not akin to the passive reflection of a mirror but rather the interaction through the mirror of the subject and the image. Catching a glimpse of myself in a mirror, I adjust my position, I change my expression, I tidy my hair and consider myself from the position of the other. As Lacan argued, by recognizing ourselves in the mirror as children, we begin to constitute ourselves as subjects (2006). Later in life, many others, including parents, colleagues, audiences and so forth hold mirrors in which we perceive ourselves and which sustain our subjectivities. As researchers, the work we carry out continuously defines and redefines us as subjects.

This has significant implications. Consider, for example, the issue of a researcher's own values and their impact on her work. A reflexive researcher does not seek to be value neutral but is alert to the ways that her research expresses, reinforces or undermines the values that she holds. To this end, she seeks transparency in the values embodied by her research in the choice of subject, choice of conceptual approaches, methodologies, field work and so forth (Myrdal 1972). Achieving transparency in the values expressed in the research we do is far from easy and calls for a close probing of our motives in doing the work we do. For example, how many of us would be prepared or willing to acknowledge that our choice of topic, of collaborators or of methodology may be dictated by careerism or by contemporary fashion? How many of us would acknowledge that our work may be driven by an ambition to publish at all costs? As reflexive researchers, we often have to work against our inner resistances and the researcher's permanent temptation to rationalize her work in terms of lofty ideals of enhancing knowledge and advancing the welfare of humanity.

Reflexivity does not only call into question the values and motives that drive our research. It also calls for a recognition of the effects of our own presence, as researchers in the field. Qualitative 'data' has the imprimatur of the researcher—it cannot claim to be objective in the way quantitative data in both natural and social sciences can. An interview or an observational session is a product of intersubjective encounters and practices, influenced by numerous psychological and circumstantial factors. The same individual observed or interviewed by different researchers will act differently and in some cases *very* differently—it is not uncommon for a respondent to say two exactly opposite things to two different researchers. I had ample opportunity to observe this when carrying out research on unemployed managers and professionals (Gabriel et al. 2010; Gabriel et al. 2013). An unemployed manager would clam up and portray himself as having a stiff upper lip, confident of getting another job shortly; the same manager, when interviewed by a different researcher, would break down in tears of despair or self-loathing and express hopelessness and despair. In a series of interviews, a younger researcher was repeatedly able to get under the defences that were raised by unemployed men and women in their 50s when interviewed by her senior colleagues. The qualitative interview is an interactive accomplishment and difficult to assess without a strong sense of the context or the occasion. In this way, the empirical material of qualitative research does not pre-exist the act of collecting it but is created in the act of research itself. It is the product of a particular type of social encounter (Gabriel 2015), one in which respondents may be seeking to impress, defer to or defy their interviewers. It is for this reason that the precise way in which researchers explain the purpose of their work to their respondents, something that is rarely mentioned in the 'methodology' sections of published papers, is of the greatest significance in contextualizing the meaning of the conversation. Snippets of an interview, which fail to mention even the gender or age of a respondent, may give the impression of supporting an often grandiloquent argument—in reality, however, all they reveal is a researcher's skill in supporting virtually any argument with appropriate, decontextualized evidence.

Acknowledging the effects of the research relation also requires a recognition of the power dimensions of this relation, something that qualitative researchers ignore at their peril (Alvesson and Sköldberg 2009). This occurs both at the individual and organizational levels. A researcher who enjoys a relatively well paid job and a comfortable life inhabits a different world from that of a social security claimant, a bereaved parent or a cancer patient. Research with so-called vulnerable groups (Becker-Blease and Freyd 2006; Booth and Booth 1994; Dyregrov 2004) has now become a regular topic of research ethics, but recognizing the asymmetries in the relation between researchers and their respondents goes beyond formulaic precautions to safeguard the rights of these groups. It calls for an acknowledgement of the emotional and symbolic dimensions of a relation that cannot leave researchers as objective observers but makes big demands on their own emotional resources.

Psychoanalytic researchers are aware that every research encounter entails some transference and countertransference, in other words, the reawakening of experiences and emotions from earlier periods of their lives in both researchers and their respondents (Czander 1993; Ulus and Gabriel 2016). Meeting somebody in the contrived environment of a guided conversation with unequal power relations frequently triggers in a respondent emotions and fantasies associated with significant figures from their past. This is known as transference and can be positive (warm and supportive feelings) or negative (fearful, envious, suspicious etc.). Countertransference represents the researchers' own response to the transference of their interlocutor, shaping our own emotions during and following an interview. A careful analysis of our countertransference can offer powerful insights into elusive unconscious emotions and help us make sense not only of the emotional dynamics of the interview situation itself but also of the deeper significance of our respondents' behaviour and utterances in our presence (Gabriel and Ulus 2015; Gemignani 2011; Stein 1999). Acknowledging our countertransference requires constant monitoring and recording of our immediate reflections and experiences during field work. For example, we may write about feeling claustrophobic, puzzled or unsettled at specific points, and such emotions can serve as additional resources for interpreting our empirical material. Countertransference, when effectively deployed, can be particularly useful in bringing to the surface unconscious dimensions in an encounter between the researcher and her respondents, which may normally be prevented from reaching the surface by various psychological defences.

The power dynamics of research are not limited to the face-to-face interaction between researchers and their subjects. As Alvesson and Sköldberg (2009) observe, every aspect of the research design and its execution entails a wide range of political and ideological assumptions which can easily go unnoticed. This is especially apparent when we revisit classic studies from the past, whether in sociology (e.g. Liebow 1967/1981; Matza 1969) or in anthropology (e.g. Malinowski 1922; Mead 2001 [1930]), and the ideological and political assumptions of their authors become immediately apparent. More specifically, qualitative research in organizations usually involves delicate negotiations for access with different powerful stakeholders. This is a delicate political game in which various undertakings may be given by researchers. Reflexive researchers question the implications of such undertakings both for the framing and execution of their work as well as for its political implications.

The political and ideological underpinnings of all social research are also present in the language used, the very terms and concepts used. Since the rise of scholarship on discourse, we have become aware that every discursive resource we deploy is laden with political assumptions. Even innocent-sounding terms like 'management', 'consumers', 'markets', 'culture', 'gender', 'stakeholders' and 'knowledge' embody powerful ideological and political assumptions from which researchers cannot detach or abstract themselves. In doing research on unemployed managers and professionals, we found that the word 'unemployed' itself carried assumptions of self-blame and self-responsibility that frequently raised emotional defences among our respondents. Using an expression like "being out of paid work" frequently drew different responses from our participants from "being unemployed". Reflexive researchers will be sensitive to these assumptions that can be notoriously difficult to disentangle and explore. Moreover, terms can carry connotations and resonances that vary from research to researcher and from researcher to the wider public. Notice for example how the long-standing convention of referring to homo sapiens through the male pronoun is now widely viewed as part of a sexist language that privileges the male of the species and frequently renders women invisible and inaudible. Notice too often apparently innocent terms appear to embody assumptions that tacitly sustain hierarchies of inclusion and privilege.

Reflexive qualitative researchers face another type of challenge. Unlike quantitative researchers in both natural and social sciences who can set their sights on confirming or disconfirming a theory or a hypothesis, qualitative researchers are often undertaking a journey whose conclusion may be very different from their original intentions; along the way of this journey, they may themselves emerge as different subjects, their ideas, methodologies and even their identities altered in some very significant ways. Reflexive researchers cannot deal with empirical material as something separate from themselves—as something stored in a computer file, to be processed, squeezed or distilled to generate knowledge at a later date. Instead, their own identity as researchers and their own practices as investigators are intimately intertwined with their experiences in the field. In writing up their research, in a thesis, an article or a monograph, qualitative researchers are thus faced with a difficult conundrum. Should they acknowledge that the destination point of their work was different from the original one, in which case their methodology was one chosen for a different purpose, or should, as they often do, create a story of how their research was always intended to reach the conclusions which they finally attained?

Several scholars discussing reflexivity in qualitative research emphasize its social nature. Reflexivity is not merely a solitary conversation with the self, but calls for a series of dialogues with multiple 'others', including audiences, research collaborators and field respondents (Mahadevan 2011; Orr and Bennett 2009). Reflexivity should not lapse into a solipsistic or narcissistic exercise (Tomkins and Eatough 2010) but should rather be a vigorous effort of questioning both individual and professional practices. Not only should the individual researcher's values and motives be interrogated but equally the professional norms of different disciplines, the practices of academic publishing and, more widely, the role of social research in society. In this connection, Alvesson et al. (2017) have criticized what they view as the loss of social relevance and meaning for a large part of research in the social sciences which has become increasingly formulaic and esoteric. Research is no longer a vocation driven by passion for learning and discovery but has become a publishing game, whereby individual researchers aim for hits in prestigious journals and the merits of different pieces of work is reduced to metrics, such as journal impact factors and citation counts. Many articles that get published amount to hardly more than footnotes to the work of others, footnotes that are only meaningful to tiny microtribes of fellow researchers, if that. Alvesson et al. castigate the explosion of published outputs which create a noisy, cluttered environment as different voices compete to capture the limelight even briefly. In an environment where the volume of published material doubles every five years or so and where most of the publications remain unnoticed and uncited, capturing the limelight, however briefly, becomes paramount. This results in widespread cynicism among many social researchers on the value of academic research including their own. Combating this type of cynicism and seeking to make their findings socially meaningful by engaging a wider public is seen as a significant challenge for reflexive researchers. This challenge is compounded by various institutional factors, including research assessment exercises, departmental rankings and the general devaluation of teaching that become internalized by academics seeking to promote their own careers and reputations.

8.4   Limits to Reflexivity

Faced with such a challenge, several voices have cautioned against viewing reflexivity as the panacea for qualitative research. Weick (2002) has criticized reflexivity as an end in itself which easily ends up as navelgazing or box-ticking rather than as a means for improving the quality of theory. The critique of reflexivity as an end in itself is shared by Rhodes (2009), who argues that reflexivity must reach beyond minuscule textual interrogation and aim towards the development of a social responsibility among researchers in the ways they exercise and promote academic freedom through their daily practices. Lynch (2000, p. 26) has been even more absolute in castigating reflexivity as an increasingly diffuse idea that serves to legitimate various types of knowledge by privileging researchers who claim to be reflexive against those who are assumed to be unreflexive. He proposes, instead, that reflexivity is "an ordinary, unremarkable and unavoidable feature of [all human] action" and should not be viewed as a special virtue and the unique preserve of those researchers who use the term as a prop for a wide variety of claims.

Such warnings are valuable. Reflexivity should certainly not be reduced to either a box-ticking exercise or solipsistic self-questioning. Nor should it be seen as a guarantee of good quality research or theorizing. All the reflexivity in the world will not turn a dull piece of work into an interesting one. Quite the opposite—it will make it still duller. What reflexivity will not replace is the researcher's intelligence and craft that are alert to similarities and exceptions, continuities and discontinuities, plans and accidents. In particular, reflexivity cannot replace two vital qualities in qualitative research—the ability to spot meaningful details and critical theoretical imagination.

Spotting significant details for a qualitative researcher is, as we saw earlier, akin to a detective's skill in identifying those tiny clues that lead to the criminal and the crime. Having an eye for the significant detail, present or missing (Schreven 2015), is the skill threatened by unquestioned reliance on coding software like nVivo to reveal the deeper meaning of a text. Observing significant details is a skill that calls not just for textual alertness, but also for a kind of vigilance. Significant details do not go unnoticed simply because they are details but because, very often, there are particular obstacles, institutional, intellectual and emotional, deliberately placed so as to hide them. Excessive preoccupation with large sizes and quantities very frequently makes us blind to the significant detail. Thus, instead of seeking to process large amounts of qualitative material as though it were quantitative, my preference is to search for small revealing details amidst large quantities of uniform, predictable and dull data. In this regard, I have suggested that qualitative researchers should act as beachcombers who survey a huge beach looking for valuable objects and materials or at least for objects and materials which offer interesting possibilities (Gabriel 2015).

A beachcomber will ignore large amounts of sand, rocks and rubbish with eyes that search restlessly for small items of value and interest. In this sense, a beachcomber is in a quest not for objects themselves but rather for the possibilities offered by different objects. To a beachcomber, a piece of driftwood may suggest things as diverse as a bonfire on the beach, an artistic installation or the existence of a nearby shipwreck. A sea-shell may suggest an addition to a mobile or may spark the inspiration for a collage or painting. In a similar way, it seems to me that an important part of a qualitative researcher's skill lies in the recognition of possibilities afforded by small details in the empirical material, details than can be easily overlooked and are unlikely to be picked up by nVivo or other coding software. Details full of interesting possibilities for qualitative researchers is exactly what Gibson (1977) calls affordances—a realm of possibilities for action opened up by different objects of the environment.

Identifying these theoretical possibilities of specific aspects of qualitative material requires that qualitative researchers go beyond the standardized routines of their methodologies and use their critical theoretical *imagination*. Imagination, a quality praised by C. Wright Mills (1959) in his book *The Sociological Imagination* that should be required reading for every qualitative researcher, is a quality present at every stage of an inquiry. It is a restless state of the mind that pressingly and persistently asks the questions "Why?", "What if?" and "So what?", one that is capable of shifting perspectives from the specific to the general and vice-versa, one that seeks theoretical openings in analogies and metaphors, one that does not seek easy closure in the face of mystery and uncertainty. Imagination is a quality that does not discard method but is willing to work 'against method' when circumstances call for it (Feyerabend 1975). Far from being the enemy of objective inquiry, imagination is an indispensable quality for challenging existing and, at times, oppressive beliefs and ideas and venturing into the realms of alternative possibilities. Imagination is an essential quality of all critical inquiry—inquiry that seeks to challenge existing assumptions (Alvesson and Sandberg 2013), common sense, comfortable routines and predictable, comfortable interpretations (Svensson 2014).

In this chapter, we have identified some core differences in the ways that quantitative and qualitative researchers go about pursuing knowledge. We have highlighted the importance of interpretation for qualitative research and have indicated some of the ambiguities and complexities researchers face in corroborating and strengthening their interpretations. We then examined the role of reflexivity as an important quality in qualitative research, one through which researchers, individually and in groups, explore the assumptions, meaning and ramifications of their own practices, as well as the wider social relevance of their contributions. As important as reflexivity is, we concluded by suggesting that reflexivity is no substitute for critical imagination—a vital quality in all qualitative research. We suggested that this quality frequently calls for a departure from the strictures of formal methodology, an approach to empirical material with a sharp eye for detail as a set of affordances rather than a uniform volume of data, to the act of interpretation itself as well as to a persistent critical deployment of the questions "Why?", "What if?" and "So what?"

References

Alvesson, M., & Sandberg, J. (2013). Has Management Studies Lost Its Way? Ideas for More Imaginative and Innovative Research. *Journal of Management Studies, 50*(1), 128–152.

Alvesson, M., & Sköldberg, K. (2009). *Reflexive Methodology : New Vistas for Qualitative Research* (2nd ed.). London: Sage.

Alvesson, M., Gabriel, Y., & Paulsen, R. (2017). *Return to Meaning: A Social Science That Has Something to Say*. Oxford: Oxford University Press.

Becker-Blease, K. A., & Freyd, J. J. (2006). Research Participants Telling the Truth About Their Lives—The Ethics of Asking and Not Asking About Abuse. *American Psychologist, 61*(3), 218–226.

Booth, T., & Booth, W. (1994). The Use of Depth Interviewing with Vulnerable Subjects: Lessons From a Research Study of Parents with Learning Difficulties. *Social Science & Medicine, 39*(3), 415–424.

Bruner, J. S. (1986). *Actual Minds, Possible Worlds.* Cambridge, MA: Harvard University Press.

Cunliffe, A. L. (2003). Reflexive Inquiry in Organizational Research: Questions and Possibilities. *Human Relations, 56*(8), 983–1003.

Czander, W. M. (1993). *The Psychodynamics of Work Organizations: Theory and Applications.* London: Guilford Press.

Dyregrov, K. (2004). Bereaved Parents' Experience of Research Participation. *Social Science & Medicine, 58*(2), 391–400.

Eisenhardt, K. M. (2002). Building Theories From Case Study Research. In A. M. Huberman & M. B. Miles (Eds.), *The Qualitative Researcher's Companion* (Vol. 50, pp. 25–32).

Feyerabend, P. (1975). *Against Method.* London: New Left Books.

Gabriel, Y. (2015). Reflexivity and Beyond: A Plea for Imagination in Qualitative Research Methodology. *Qualitative Research in Organizations and Management: An International Journal, 10*(4), 332–336.

Gabriel, Y., & Ulus, E. (2015). "It's All in the Plot": Narrative Explorations of Work-Related Emotions. In H. Flam & J. Kleres (Eds.), *Methods of Exploring Emotions* (pp. 36–45). Abingdon: Routledge.

Gabriel, Y., Gray, D. E., & Goregaokar, H. (2010). Temporary Derailment or the End of the Line? Managers Coping with Unemployment at 50. *Organization Studies, 31*(12), 1687–1712.

Gabriel, Y., Gray, D. E., & Goregaokar, H. (2013). Job Loss and Its Aftermath Among Managers and Professionals: Wounded, Fragmented and Flexible. *Work, Employment & Society, 27*(1), 56–72.

Gadamer, H.-G. (1975). Hermeneutics and Social Science. *Cultural Hermeneutics, 2*(4), 307–316.

Gemignani, M. (2011). Between Researcher and Researched: An Introduction to Countertransference in Qualitative Inquiry. *Qualitative Inquiry, 17*(8), 701–708.

Gibson, J. J. (1977). The Theory of Affordances. In R. Shaw & J. Bransford (Eds.), *Perceiving, Acting, and Knowing: Toward an Ecological Psychology* (pp. 67–82). Mahwah: Lawrence Erlbaum.

Ginzburg, C. (1980). Morelli, Freud and Sherlock Holmes: Clues and Scientific Method. *History Workshop, 9*, 5–36.

Habermas, J. (1972). *Knowledge and Human Interests*. London: Heinemann.

Hardy, C., Phillips, N., & Clegg, S. (2001). Reflexivity in Organization and Management Theory: A Study of the Production of the Research 'Subject'. *Human Relations, 54*(5), 531–559.

Hibbert, P., Coupland, C., & MacIntosh, R. (2006). Reflexivity: Recursion and Relationality in Organizational Research Processes. *Qualitative Research in Organizations and Management: An International Journal, 5*(1), 47–62.

Lacan, J. (2006). The Mirror Stage as Formative of the Function of the I as Revealed in Psychoanalytic Experience. In J. Storey (Ed.), *Cultural Theory and Popular Culture: A Reader* (pp. 287–292). London: Pearson.

Liebow, E. (1967/1981). *Tally's Corner: A Study of Negro Streetcorner Men*. Boston: Little, Brown and Company.

Lynch, M. (2000). Against Reflexivity as an Academic Virtue and Source of Privileged Knowledge. *Theory, Culture and Society, 17*(3), 26–54.

Mahadevan, J. (2011). Reflexive Guidelines for Writing Organizational Culture. *Qualitative Research in Organizations and Management: An International Journal, 6*(2), 150–170.

Malaurent, J., & Avison, D. (2017). Reflexivity: A Third Essential 'R' to Enhance Interpretive Field Studies. *Information & Management*. http://www.science-direct.com/science/article/pii/S0378720617300629?via%3Dihub

Malinowski, B. (1922). *Argonauts of the Western Pacific*. London: G. Routledge & Sons.

Matza, D. (1969). *Becoming Deviant*. Englewood Cliffs: Prentice-Hall.

Mead, M. (2001 [1930]). *Coming of Age in Samoa: A Psychological Study of Primitive Youth for Western Civilisation*. New York: Perennial Classics.

Mills, C. W. (1959). *The Sociological Imagination*. Harmondsworth: Penguin.

Myrdal, G. (1972). The Place of Values in Social Policy. *Journal of Social Policy, 1*(1), 1–14.

Orr, K., & Bennett, M. (2009). Reflexivity in the Co-production of Academic-Practitioner Research. *Qualitative Research in Organizations and Management: An International Journal, 4*(1), 85–102.

Rhodes, C. (2009). After Reflexivity: Ethics, Freedom and the Writing of Organization Studies. *Organization Studies, 30*(6), 653–672.

Schön, D. A. (1983). *The Reflective Practitioner : How Professionals Think in Action*. New York: Basic Books.

Schreven, S. (2015). On the Case of the Missing Detail and the Twisted Truth About Hard Work. *Organization, 22*(5), 702–719.

Stein, H. F. (1999). Countertransference and Understanding Workplace Cataclysm: Intersubjective Knowledge and Interdisciplinary Applied Anthropology. *High Plains Applied Anthropologist, 19*(1), 10–20.

Svensson, P. (2014). Thickening Thick Descriptions: Overinterpretations in Critical Organizational Ethnography. In E. Jeanes & T. Huzzard (Eds.), *Critical Management Research: Reflections From the Field* (pp. 173–188). London: Sage.

Taylor, C. (1971). Interpretation and the Sciences of Man. *Review of Metaphysics, 25*(1), 3–51.

Tomkins, L., & Eatough, V. (2010). Towards an Integrative Reflexivity in Organisational Research. *Qualitative Research in Organizations and Management: An International Journal, 5*(2), 162–181.

Tsoukas, H. (1992). Postmodernism, Reflexive Rationalism and Organizational Studies: A Reply to Martin Parker. *Organization Studies, 13*(4), 643–649.

Ulus, E., & Gabriel, Y. (2016). Bridging the Contradictions of Social Constructionism and Psychoanalysis in a Study of Workplace Emotions in India. *Culture and Organization*, 1–23. http://www.tandfonline.com/doi/full/10.1080/14759551.2015.1131688

Weick, K. E. (2002). Essai: Real-Time Reflexivity: Prods to Reflection. *Organization Studies, 23*(6), 893–898.

Willig, C. (2001). *Introducing Qualitative Research in Psychology: Adventures in Theory and Method*. Buckingham: Open University Press.

9

The Emotional Nature of Qualitative Research

Angela Stephanie Mazzetti

9.1 Introduction

This chapter aims to explore an issue that is rarely discussed in management research 'how-to' textbooks: the emotional nature of qualitative research. This chapter comprises five sections. In Sect. 9.2, I present an overview of emotions and explain some of the key terms used throughout the chapter. In Sect. 9.3, I discuss the emotional challenges of existing in two worlds: our own world and our participants' world. In Sect. 9.4, I explore some of the emotional challenges of developing rapport with research participants. I also discuss the difficulties of conducting research with our friends and conducting research in our own employing organisation. In Sect. 9.5, I explore the complexities of discussing sensitive topics either by design or by accident. In each section, I draw on a carefully selected literature set to explain the key concepts, and I present

A.S. Mazzetti (✉)
Queen's University Belfast, Belfast, UK

© The Author(s) 2018
M. Ciesielska, D. Jemielniak (eds.), *Qualitative Methodologies in Organization Studies*,
https://doi.org/10.1007/978-3-319-65217-7_9

examples from qualitative research studies as an illustration of these concepts. Finally, in Sect. 9.6, I present a series of suggestions to help prepare you to engage with the emotions.

9.2   Understanding Emotion

Our emotions influence much of what we do and how we do it and are an essential part of being 'human' (Lazarus 1991). Emotions are the complex reactions we have to our judgements about ourselves in the world, influenced by our sense of who we *feel* we are and who we want to be (Lazarus 1999). Our emotions are therefore ignited when we encounter situations that either threaten or enhance our sense of self, our values and beliefs, and our aspirations and goals (Lazarus and Folkman 1984). We experience negative emotions when we encounter situations that have the potential to threaten our personal values and aspirations or harm our well-being, and positive emotions when we encounter situations that contribute to the achievement of our ambitions and enable us to flourish (Lazarus 1991). Emotions can be conceptualised as both 'inner' and 'outer' states and, as such, we *feel* our emotion inwardly, but we also display our emotion outwardly to others and these outward displays are regulated by the cultural and social worlds to which we belong (Svašek 2005). Hochschild (1979, p. 560) uses the term 'emotion work' to refer to "the act of trying to change in degree or quality an emotion or feeling", suggesting that in social encounters, we engage in the management of our emotions to either evoke an emotion that is absent or suppress an emotion that is present. She noted that 'feeling rules' determine the culturally scripted norms and expectations about how we *should* feel and act in given situations and therefore which emotions *should* be absent or present. Hochschild (1983) notes that feeling rules also determine the appropriate 'bows from the heart', the appropriate emotional exchanges, such as facial expressions, choice of words or tone of voice expected by others to convey to them that we are feeling the appropriate emotion. However, she notes that feeling rules are culturally scripted, creating a dilemma as to how 'much' emotion we display; too much emotion and we may come across as insincere, but too little and we may appear cold and detached (Hochschild 1983).

Qualitative studies are, by their nature, emotional, as they involve human researchers interacting with human participants (Broussine et al. 2014); interactions which ignite our feelings and emotions about ourselves and our participants (Clarke and Knights 2014). However, although the emotional nature of conducting qualitative research is openly discussed in other disciplines (see Dickson-Swift et al. (2008) for a review in health research and Milton and Svašek(2005) for a review in anthropological research), within business and management studies, there has been limited reference to the issue (Broussine et al. 2014; Clarke et al. 2014; Mazzetti 2013; Whiteman et al. 2009). Yet, the research process is "marinated in all kinds of emotions" (Broussine et al. 2014, p. 11). A number of researchers have suggested that the lack of discussion regarding the emotions stems from an academic community dominated by positivism, objectivity and science, (Cassell and Symon 2012; Watts 2014) and, as such, the rational is still privileged over the emotional (Brannan 2014). This has resulted in the plethora of 'how-to' textbooks focusing on technique rather than on skills with the emotional and embodied aspects of conducting qualitative studies rarely explored (Clarke and Knights 2014). However, not addressing the emotional nature of qualitative research in organisational settings risks researchers being unprepared for their emotional encounters (Whiteman et al. 2009) and may result in researchers feeling inadequate or incompetent (Clarke et al. 2014; Mazzetti 2013). Yet, if we embark on a study of human lives, we have to be ready to face these human feelings (Hoffmann 2007). This chapter aims to address this gap by providing some advice and guidance on how to be prepared for our own and our participants' emotions.

9.3   Straddling Two Worlds

When we embark on a research project, we are motivated by thinking that our research is important and will have a positive impact for both our research participants and the academic community (Down et al. 2006). However, as we enter a research setting, we find ourselves straddling two social worlds—our own and our participants'—and we walk a fine line between closeness (with our own world) and distance (with our participants' worlds) (Agar 1980; Down et al. 2006). These worlds are bounded by cultural, social and organisational 'rules' (Down et al. 2006; Whiteman et al. 2009) and therefore our interactions with our participants are founded on varying degrees of 'strangeness' and 'familiarity' with these 'rules' (Tietze 2012). We often find that our research participants are not as enthusiastic about our research intentions as we are, and in some cases, they can be indifferent, dismissive and resentful, or misinterpret the purpose of our research and therefore our role as a researcher (Brannan 2014; Down et al. 2006). This can create emotional dissonance as we experience challenges to the legitimacy of our research and therefore threats to our researcher identity (Down et al. 2006).

Brannan (2014) highlights that we engage in the academic inquiry of a topic that is situated within a defined academic conversation; however, he notes that when we engage with our research setting, these academic 'frames of reference' become 'operationalised' and as such the focus of our research may not necessarily be perceived in the same way by our participants. He illustrates this point by making reference to his own ethnographic study in a call centre, noting that despite his candour as to the intent of his research (he was researching employee behaviour), it was often misunderstood or misinterpreted by participants (who considered that he was researching knowledge exchange or teamworking), highlighting that although an academic community may see clear boundaries and distinctions between different areas of academic inquiry, our research participants may not. Brannan (2014) adds that misinterpretations as to the purpose of his research made him feel vulnerable as an academic and he started to question the 'legitimacy and usefulness' of his research, to question if his research design was flawed and to question if his research questions were misguided. He feared that if he was unable to clearly articulate the purpose of his research to his participants, how could he do so to a more hostile academic community?

Misunderstanding as to the purpose of our research can also create tensions between us and our participants. In organisational settings, researchers are not always welcomed, as those who grant access for the research to be conducted may not always be those who are directly interacting with the researcher (Brewer 2000) and, in certain contexts, for example, organisational ethnographies, participants may be less able to 'opt out' of the research process (Hammersley and Atkinson 2007). This may lead to feelings of suspicion and resentment towards a researcher who asks questions and records responses 'on behalf of the management' (Brewer 2000). For example, Knights (cited in Clarke and Knights 2014) refers to a difficult interview with a union representative during his research at a manufacturing plant. This participant interpreted the purpose of the research solely as a means of getting information for the 'management' and, as such, he refused to allow the interview to be recorded and his responses were monosyllabic and purposefully guarded and limited. Researchers can therefore often feel like 'interlopers' and 'outsiders' (Clarke and Knights 2014). Indeed, Garrety (cited in Down et al. 2006, p. 95) notes that during her observations of management meetings at a steel works, she felt "out of place, like a spy or a parasite".

Clarke and Knights (2014) note that interactions with participants will also be informed by attitudes to our physical identity in terms of age, gender, class, sexuality and race. Clarke (cited in Clarke and Knights 2014) notes that she felt 'apprehension and trepidation' during her research with male engineers, as she was so very different from those she was researching; she was neither male nor an engineer. Garrety (cited in Down et al. 2006, p. 95) notes her discomfort working in a male-dominated environment explaining that "although I tried to be friendly, I found it difficult (impossible!) to do the matey blokey stuff. I thought it would be too patently false to even try". However, Clarke notes that because she was different, her male engineers felt liberated to tell her things that they would not normally discuss in their male-dominated environment. Because she was female, they considered her a 'safe pair of hands' in which they could talk about their feelings and sentiments in a way that they could not normally do with their peers.

Clarke and Knights (2014) further highlight several challenges facing young researchers who may be perceived as inexperienced and naive. They suggest that young researchers may have their research manipulated by organisational gatekeepers and are therefore more likely to make ill-informed decisions due to lack of experience. Knights refers to his own research with a newspaper sales organisation, where the release of his preliminary findings to the workforce (by order of the management but unbeknownst to him) led to the staff engineering his removal from the organisation. The lesson of this story, he suggests, is that researchers should never underestimate the power of organisational politics. Clarke and Knights (2014, p. 48) therefore highlight that "your identity as a researcher is not so much about your idea of who you are but the potential myriad of interpretations and projections made by your audience".

9.4   Developing Relationships

An important aspect of qualitative research is developing 'rapport' with our research participants, and to do this, we need to be able to demonstrate an understanding of our participants' worlds, an interest in their stories and empathy with their feelings (Dickson-Swift et al. 2008). In their research of the challenges encountered by PhD students undertaking qualitative interviews, Roulston et al. (2003) highlight that new researchers are often anxious about dealing with unexpected participant behaviours, dealing with their own actions and biases, keeping the interview flowing and focused and dealing with sensitive issues. They further highlighted a number of practical issues that can impact the researcher/participant interaction, including dealing with distractions such as noise, dealing with eavesdroppers, taking notes during discussions and dealing with the 'stumbles' and 'slips' of normal conversations. Qualitative researchers therefore need to be able to outwardly display their interest, understanding and empathy with their participants' stories but within the appropriate culturally scripted levels of exchange (Dickson-Swift et al. 2008; Hoffmann 2007).

Not all of our encounters with participants will be positive ones. Sometimes, we meet people with whom we struggle to develop a rapport (Down et al. 2006; Kleinman and Copp 1993) and, at times, we may encounter those who are hostile or rude towards us despite our best efforts to engage with them (Dundon and Ryan 2010; Mazzetti 2013). The on-going effort of trying to establish a rapport (when there clearly is not one to be established) requires not only significant emotion work on behalf of the researcher (Dickson-Swift et al. 2008) but also a need for us to engage in alternative strategies to find some connection with our participants (Dundon and Ryan 2010). Sometimes, we may avoid engaging with those we do not like, but in doing this, we risk potentially missing out on hearing different or alternative perspectives (Down et al. 2006; Kleinman 1991). In some encounters, we may need to suppress the negative emotions we feel for our participants (Down et al. 2006), which means we have to 'hang onto' all that emotion until we have an opportunity for a safe release (Dickson-Swift et al. 2008). It is important that there is a safe outlet for these emotions, for example by recording our feelings in a reflexive diary or by discussing our feelings with our research supervisor or a research buddy (Mazzetti 2013), as failing to address the issue may impact our long-term engagement with qualitative research or result in us becoming disenfranchised with our research (Kleinman 1991).

Perriton (2000) suggests that new researchers often engage with peer groups, family, friends and even partners in search of a 'research nirvana', anticipating that it will be easier to research those with whom there is already an established relationship rather than trying to build a relationship with a 'stranger'. However, she adds that just because there is an established relationship does not mean that the difficulties inherent in research with 'strangers' are absent. Indeed, she notes that this new relationship of friend/researcher and friend/participant often results in feelings of embarrassment and 'bewilderment', suggesting that friendship and research are a 'potent mix'. She illustrates with an example of a rather uncomfortable 'research' lunch with a friend during which the normal lunchtime protocols such as where to go and who should pay became rather strained and the whole conversation had an atmosphere of tension not normally experienced under normal lunchtime discussions.

Additionally, many researchers conduct their research in their own (employing) organisation. Tietze (2012) and Perriton (2000) suggest that the growth in professional qualifications which require the completion of a dissertation, project or thesis has resulted in an increase in such 'insider researcher' studies (Brannick and Coghlan 2007). A number of advantages and disadvantages to 'insider research' have been put forward. Tietze (2012, p. 56) notes that such close involvement and familiarity with the research setting, can ensure "access to organizational niches and nooks, which may remain hidden otherwise". However, Brannick and Coghlan (2007) note that some avenues may be closed off to insider researchers because of their position in the organisation's hierarchy: too low in the hierarchy and insider researchers may find that they cannot gain access to the more senior networks; too high in the hierarchy and they may not get access to the informal and grapevine networks. Additionally, these different 'actors' in the system may hold very different expectations and suspicions about the purpose of the research (Grisoni and Broussine 2014).

Insider researchers take on multiple roles as they become "researchers as well as employees, managers, bosses, occupational experts, colleagues and friends" (Tietze 2012, p. 59) and, as such, they are likely to experience role conflict as they are caught between the sometimes conflicting demands of their organisational role(s) and their researcher role (Brannick and Coghlan 2007; Grisoni and Broussine 2014). Tietze (2012) suggests that the 'researcher' and the 'researched' may have conflicting views as to what it means to be a 'colleague' and a 'researcher participant' at the same time, putting a strain on relationships, a strain which may extend beyond the life of the research project. Referring to her own PhD 'insider' research of student and staff in an educational setting, Grisoni (cited in Grisoni and Broussine 2014) highlights the challenges she encountered taking on the roles of line manager, researcher and tutor. She notes that being an insider had advantages in terms of preknowledge and an understanding of the research context, but also had implications for the power relationships between her and her participants. She suggests that, at times, she would 'shy away' from asking challenging or follow-up questions during interviews due to the conflicting roles of line manager and researcher. Indeed, Brannick and Coghlan (2007) highlight that our employing organisation may put constraints on what can be disclosed and documented leaving us with the dilemma of 'report' (and live with the aftermath) or 'don't report' (and keep our job). Grisoni (cited in Grisoni and Broussine 2014) also highlights the practical difficulties in switching between the role of 'researcher' and that of 'employee', noting that she would often rush through a research meeting in order to prepare for a work meeting. She learned to allocate days specifically for research so that she was not squeezing research activities in between work activities, which she suggests gave her a clearer research focus.

9.5   Dealing with Emotive Topics

Hoffmann (2007) suggests that the focus of our research may be sensitive and therefore emotive. Lee (1993) defines sensitive research as being intrusive in nature (e.g. the study of private or personal stories and issues), incriminating (e.g. the study of deviant behaviours) or political (e.g. the study of power imbalances or coercion). For example, Hoffmann (2007) suggests that her research of 'problems in the workplace' was by its nature an emotive topic for many of the participants. She discusses the emotion work required to demonstrate just the right amount of empathy with her participants but without being too emotional or too detached. She also highlights a common problem for many management researchers, that of being caught in the middle between managers and subordinates and therefore having heard 'the story' from both perspectives. She illustrates this point by referring to an interview with a manager who became emotional when discussing how his subordinates perceived him as a manager. She highlights that she was unable to console him by saying "that's not true, they don't think of you in that way" (because she knew they did) or defend him by saying "don't worry about what they think" (because this would have been dismissive of their views) and, as such, we can find ourselves in situations where we can neither 'soothe or defend' our respondents.

However, Lee (1993) suggests that what is considered 'sensitive' is 'highly contextual' as even fairly innocuous topics can become highly emotive because of the situational context in which they are researched. He refers to John Brewer's (1991) ethnographic study of routine policing (a fairly innocuous topic), which was highly emotive as a result of the social conflict in Northern Ireland at the time of the study. Lee (1993) therefore suggests that it may not necessarily be the topic *per se* that is sensitive but rather the conditions under which sensitivity may arise. For example, my own ethnographic study with the fire and rescue service took place during a period of tense industrial relations, government reforms, spending cuts, and industrial action. This created a situational sensitivity to my research I had not anticipated (Mazzetti 2016).

Roulston et al. (2003) highlight that a key anxiety for new researchers undertaking qualitative interviews is that although we might anticipate our conversations to follow a particular format, we can never be sure what way the interview will progress and the topics we will end up discussing. Indeed, the emergent and participant-led nature of qualitative research means that we often find ourselves discussing topics that we had not anticipated (Hoffmann 2007). This means that we can often 'trespass' into sensitive territory (Lee 1993). Roulston et al. (2003) note that new researchers fear discussing sensitive topics as interviews can 'get too tough'. They suggest that this influences the researcher's ability and willingness to ask probing questions. However, Van Maanen (2010, p. 338) states that we should not shy away from such issues, suggesting that "without an affront, injustice, complaint, or beef to explore we might well become ciphers-qua-celebrants, happy agreeable sorts who wallow in unmitigated delight within the organizations we study and, in the end, have little to say other than everything is hunky-dory".

9.6   Conclusions

Following on from the discussion and examples presented in this chapter, I have suggested some ideas to help prepare you for emotional encounters.

- Be aware of your own biases and 'grievances' and reflect on how these can influence your choice of research topic, your choice of research setting, your research design, how you engage with participants and how you interpret and present your data.
- Read the papers and books cited in this chapter as they provide detailed academic debates and more practical advice and guidance on the topic of researcher emotion. Reflect on the issues they raise and prepare an action plan for your own development.
- Talk to your research supervisor about any worries and concerns you may have about your research and also about any emotional encounters you may experience. It is important that you have a safe outlet for such discussions. Also, find out if your institution offers any other formal support mechanisms for researchers.

- Find out if your institution has a research buddy system or researcher action learning set. These peer support mechanisms can often provide an additional and more informal source of advice and guidance.
- Keep a reflective diary to record your feelings and emotions.
- If you are new to qualitative research, take advantage of training and development activities aimed at helping to develop your researcher skills such as interviewing techniques. It is safer to hone such skills in the safe environment of a training room.
- Accept that we all have research encounters that we wish had gone better. Reflect on what happened, discuss it with your supervisor or research buddy, and then action plan for future development.

References

Agar, M. (1980). *The Professional Stranger: An Informal Introduction to Ethnography*. New York: Academic Press.

Brannan, M. J. (2014). Recognising Research as an Emotional Journey. In C. Clarke & M. B. a. L. Watts (Eds.), *Researching with Feeling: The Emotional Aspects of Social and Organizational Research* (pp. 17–34). London: Routledge.

Brannick, T., & Coghlan, D. (2007). In Defense of Being "Native": The Case for Insider Academic Research. *Organizational Research Methods, 10*, 59–74.

Brewer, J. (1991). *Inside the RUC: Routine Policing in a Dived Society*. Oxford: Clarendon Press.

Brewer, J. (2000). *Ethnography*. Berkshire: Open University Press.

Broussine, M., Watts, L., & Clarke, C. (2014). Why Should Researchers Be Interested in Their Feelings? In C. Clarke, M. Broussine, & L. Watts (Eds.), *Researching with Feeling: The Emotional Aspects of Social and Organizational Research* (pp. 1–16). London: Routledge.

Cassell, C., & Symon, G. (2012). Introduction: The Context of Qualitative Organizational Research. In G. Symon & C. Cassel (Eds.), *Qualitative Organizational Research* (pp. 1–11). London: Sage.

Clarke, C., Brousinne, M., & Watts, L. (2014). *Researching with Feeling: The Emotional Aspects of Social and Organizational Research*. London: Routledge.

Clarke, C., & Knights, D. (2014). Negotiating Identities. In C. Clarke, M. Broussine, & L. Watts (Eds.), *Researching with Feeling: The Emotional Aspects of Social and Organizational Research* (pp. 35–50). London: Routledge.

Dickson-Swift, V., James, E., & Liamputtong, P. (2008). *Undertaking Sensitive Research in Health and Social Sciences: Managing Boundaries, Emotions and Risks*. Cambridge: Cambridge University Press.

Down, S., Garrety, K., & Badham, R. (2006). Fear and Loathing in the Field: Emotional Dissonance and Identity Work in Ethnographic Research. *Management, 9*, 87–107.

Dundon, T., & Ryan, P. (2010). Interviewing Reluctant Respondents: Strikes, Henchmen, and Gaelic Games. *Organizational Research Methods, 13*, 562–581.

Grisoni, L., & Broussine, M. (2014). Emotionally Charged Research. In C. Clarke, M. Broussine, & L. Watts (Eds.), *Researching with Feeling: The Emotional Aspects of Social and Organizational Research* (pp. 51–65). London: Routledge.

Hammersley, M., & Atkinson, P. (2007). *Ethnography*. King's Lynn: MPG Books Group.

Hochschild, A. (1979). Emotion Work, Feeling Rules, and Social Structure. *American Journal of Sociology, 85*, 551–575.

Hochschild, A. (1983). *The Managed Heart: Commercialisation of Human Feeling*. Berkeley: University of California Press.

Hoffmann, E. (2007). Open-Ended Interviews, Power and Emotional Labour. *Journal of Contemporary Ethnography, 36*, 318–346.

Kleinman, S. (1991). Field-Workers' Feelings: What We Feel, Who We Are, How We Analyse. In W. B. Shaffir & R. Stebbins (Eds.), *Experiencing Fieldwork: An Inside View of Qualitative Research* (pp. 184–195). London: Sage.

Kleinman, S., & Copp, M. (1993). *Emotions and Fieldwork, Qualitative Research Methods Series 28*. London: Sage.

Lazarus, R. S. (1991). *Emotion and Adaptation*. New York: Oxford University Press.

Lazarus, R. S. (1999). *Stress and Emotion*. New York: Springer Publishing Company.

Lazarus, R. S., & Folkman, S. (1984). *Stress Appraisal and Coping*. New York: Springer Publishing Company.

Lee, R. M. (1993). *Doing Research on Sensitive Topics*. London: Sage.

Mazzetti, A. S. (2013). Occupational Stress Research: Considering the Emotional Impact for the Qualitative Researcher. *Research in Occupational Stress and Well Being: Role of Emotion and Emotional Regulation, 11*, 287–313.

Mazzetti, A. S. (2016). An Exploration of the Emotional Impact of Organisational Ethnography. *Journal of Organizational Ethnography, 5*, 304–316.

Perriton, L. (2000). Incestuous Fields: Management Research, Emotion and Data Analysis. *Sociological Research Online, 5*, 1–12.

Roulston, K., de Marrais, K., & Lewis, J. B. (2003). Learning to Interview in the Social Sciences. *Qualitative Inquiry, 9*, 643–668.

Svašek, M. (2005). Introduction: Emotions in Anthropology. In K. Milton & M. Svašek (Eds.), *Mixed Emotions: Anthropological Studies of Feeling* (pp. 1–23). Oxford: Berg.

Tietze, S. (2012). Researching Your Own Organisation. In G. Symon & C. Cassell (Eds.), *Qualitative Organizational Research* (pp. 54–71). London: Sage.

Van Maanen, J. (2010). You Gotta Have a Grievance: Locating Heartbreak in Ethnography. *Journal of Management Inquiry, 19*, 338–341.

Watts, L. (2014). A Psychosocial Approach to Researching with Feeling. In C. Clarke, M. Broussine, & L. Watts (Eds.), *Researching with Feeling: The Emotional Aspects of Social and Organizational Research* (pp. 79–91). London: Routledge.

Whiteman, G., Mu¨ller, T., & Johnson, J. M. (2009). Strong Emotions at Work. *Qualitative Research in Organizations and Management: An International Journal, 4*, 46–61.

10

Accessible Research: Lowering Barriers to Participation

Daniela Rudloff

10.1   Introduction

Regardless of the specific epistemological approach, wanting to capture the lived experience of the individual is at the very heart of conducting qualitative research and often this is explicitly acknowledged by talking about 'making voices heard' (e.g., Postle et al. 2005; Cheyns 2014). But how do we help making voices heard? Two conditions need to be met: researchers need to (1) listen and (2) allow participants to speak up. And yet, research methods do not give everyone the same opportunity to speak up and marginalised groups are likely to be further excluded from research. In some cases, this is because of an unwillingness or inability to reach out to vulnerable communities (Reid 1993; McDonald and Keys 2008); in other cases, reaching out is only the first step—making it possible for people to actually participate another is the second step.

D. Rudloff (✉)
University of Leicester, Leicester, UK

© The Author(s) 2018
M. Ciesielska, D. Jemielniak (eds.), *Qualitative Methodologies in Organization Studies*,
https://doi.org/10.1007/978-3-319-65217-7_10

10.2   Representativeness and Representation

In 2014, the Office for National Statistics (ONS) estimated that in Great Britain 'there are over 11 million people with a limiting long term illness, impairment or disability' (ONS 2014). The prevalence of disability increases with age, from 6% for children to 45% for people at State Pension age (ONS), although the prevalence of disabilities is likely to be underreported and the actual rates therefore higher (Abberley 1992).

In the ONS report, the term 'disability' covers a wide range of sensory and mobility impairments as well as mental health illnesses, learning disabilities and non-neurotypicality; consequently, there is a wide range of impairments which can lead people to encounter different barriers to participation and do so to different degrees. Statistically speaking, there is a high likelihood that you have encountered or will encounter disabled participants in your past, current or future research: Even at the lowest reported level of 6%, the prevalence established for disabilities in children, and not accounting for corrections for presumed underreporting, in a sample of 35 participants, two would be considered disabled under the ONS guidelines; when conducting research with participants past their retirement age, that number rises to 2 out of 5, or 15 out of 35. Of course, these are the numbers *assuming the sample is representative*. But for research to be representative, participation must be *possible*, and for participation to be possible, methods need to be *accessible*. However, there is increasing awareness that research methods can present barriers to participation (Wilson et al. 2013; Whitney 2006), ranging from phone surveys which are barely or not at all accessible to hearing-impaired or deaf people; household surveys excluding people with mobility issues who do not answer the door fast enough (both from Parsons et al. 2000); to focus groups putting substantial strain on elderly participants (Barrett and Kirk 2000), to name just a few examples.

In some of these cases, research participation can be literally inaccessible because participation is either made difficult or impossible. Research can impose further barriers by not being inclusive wherever assumptions about the lived experience of (potential) participants are expressed, such as regarding their gender, sexual orientation, faith, and their health and communication abilities.

Despite a growing body of research on the multiple barriers disabled people can encounter in their day-to-day lives (e.g., Easton 2013; Hammel et al. 2015), and a strong movement towards making research more accessible for participants with cognitive disabilities (e.g., Nind and Vinha 2012; Bigby et al. 2014), accessibility is often neglected as a consideration in the design of research methods. There are few or no formal 'checkpoints' which would require a researcher to explicitly reflect or report on accessibility considerations, whether for funding or for dissemination. For example, there is no mention of the requirements of making research accessible except in the context of using 'accessible language' and making sure to safeguard vulnerable participants (almost exclusively defined as participants with cognitive disabilities) in the ethics guidelines of major research and funding organisations, such as the Economic and Social Research Council (ESRC 2010), British Academy of Management (BAM 2013), British Sociological Association (BSA 2006) or the Wellcome Trust (Wellcome Trust 2017). Similarly, few journals (outside research areas such as *Disability Studies* and *Medical Research*) require authors to report on any considerations of accessibility.

At the same time, measuring the extent of the resulting exclusion is difficult since any systematic exclusion would only be observable by comparing the relevant demographic data with the overall population demographics but, of course, this requires *collecting* the relevant demographic data in the first place, and unlike age, gender and education, information on disability and impairments is not usually collected *unless* the research question is already focussed on disability.

The consequences of unequal access to research participation are stark. Farmer and Macleod (2011, p. 6, emphasis mine) warn that '[f]ailing to involve disabled people in *research that concerns their experiences* or to communicate research results in an accessible way could mean that research alienates disabled research participants'. The limitation to 'research that concerns their experiences' is interesting to note since it would have been equally, or more, valid to simply state 'failing to involve disabled people in research could mean that research alienates disabled research participants'. I similarly take issue with their advice that 'the appropriate level of involvement, and the extent to which you ensure data collection is accessible to disabled people, should remain in line with the objectives and the likelihood of disabled people falling within the sample' (p. 5); instead, I would argue that research needs to be accessible on principle; more so, since any such estimate would be based on what are likely underreported levels of prevalence (Abberley 1992).

This exclusion may not be intentional but could be the result of a lack of awareness or reluctance to face additional difficulties:

> Social exclusion in research can be the result of investigators not being 'aware' or not choosing to look into the direction of the excluded (Kroll 2011). Participants are thought of as being 'too difficult' or 'too hard to find'. Investigators may cite tight study budgets, resource and time scarcity that do not allow for extensive accommodations. (Kroll 2011, p. 66)

The resulting skew may very well be substantial. We know that in some aspects, disabled employees experience the workplace differently to their non-disabled colleagues (e.g., Lockyer 2015; Jones 2016; Mik-Meyer 2016) and present different values relating to work and the workplace (e.g., Ville and Ravaud 1998). Impairments and disability can also affect how disabled employees' gender is perceived (e.g., Mik-Meyer 2015). Any qualitative research on people's experiences in the workplace would therefore be lacking crucial facets of lived experience.

To deny participation by failing to improve access thus likely means perpetuating existing structures of inequality at the multiple intersections of, for example (dis)ability and gender (Mik-Meyer 2015; Kavanagh et al. 2015); race, gender and disability (Petersen 2012; McDonald et al. 2007); or class and disability (Petersen 2012; Nind 2008), leading to the poignantly named 'body' of knowledge being dominated by the experiences of non-disabled participants.

This chapter therefore aims to make you more aware of potential barriers to participation; to help you remove or lower those barriers so that more people can participate; and to subsequently give more people a voice, making your research more representative of the population in general. In doing so, I am keenly aware of my own privilege as a non-disabled, white woman. I am also aware that my role and position as an academic affords me further relative privilege to make my own voice heard, and to do so from relative financial safety and security. My aim is to use this privilege afforded to me in allyship to marginalised groups in the hopes that this is seen as such, and neither as exploitation nor condescension or appropriation. Any mistakes made, any information omitted, any resultant need for education, and any offense caused is no one's fault or responsibility but my own.

A final caveat: This chapter relates to research that does not explicitly focus on disability, and, in doing so, further focuses on the implications of (potential) participants with physical impairments, such as, for example sensory and mobility impairments. The exclusion of mental health, cognitive impairments, learning disabilities and non-neurotypicality should not be understood as an expression of importance or priority, but reflects the limited scope a single chapter can necessarily cover.

10.3 The Accessible Research Process

Improving access to participation starts well in advance of the first contact with a participant, and it extends beyond data collection to data analysis and dissemination. The following sections will look at individual elements of the research process, from the planning stage, to recruiting and communicating with participants, the main stage of qualitative data collection, and to the dissemination of research findings.

10.3.1 Planning Accessible Research

Your research question and your epistemology determine the major framework, methodology and method of your research; yet, you likely will have some degree of freedom in how you design, realise and embody the actual research method. Different methods can put different groups at a (dis)advantage and methods which may improve access for some may make it more difficult to participate for others: For example, Schroedel (1984) established that hearing-impaired or deaf individuals preferred face-to-face interviews, as this provided additional visual cues to understanding; Farmer and Macleod (2011) similarly recommend face-to-face or, alternatively, written or postal communication for hearing-impaired participants. Yet, printed material and postal communication can put participants with a visual impairment at a disadvantage. Parsons et al. (2000) therefore recommend mixed methods or multiple methods to increase participants' chances to find a mode of participation that suits their communication needs. Farmer and Macleod (2011, p. 36) similarly suggest using in-depth interviews or ethnography over surveys because of the flexibility this provides for participants with learning disabilities, communication issues or neuro-diversity impairments, adding that 'for people who use specific tools to enable communication, such as Talking Mats, it would also be appropriate to develop a methodology that allows them to respond using their preferred method of communication'.

The planning stage also includes budgeting for potential extra costs which could improve accessibility, such as costs for a sign language interpreter, providing Braille versions of documents, testing and improving accessibility of online tools, or budgeting for more and smaller groups. You may find that some of these possible accommodations are less expensive and therefore more feasible, whereas others may appear too expensive and therefore impractical. For example, hourly rates for a fully qualified Sign Language Interpreter start at £70/hour plus a call-out fee (National Union of British Sign Language Interpreters 2017); hiring interpreters for a series of interviews or focus group sessions can therefore quickly become a noticeable cost in a budget. That said, funders cannot grant funds for which they are not asked; university departments cannot budget for costs no one requests and argues for.

10.3.2 Approaching and Recruiting Participants

Communication in the recruitment phase is particularly important since it is at this stage that individuals become (potential) participants. Whenever you do not communicate possible accommodations in advance, your potential participants will have to explicitly ask for accommodations, which is putting the onus on them—or take a risk by turning up and finding out whether participation is possible. This essentially asks potential participants to perform emotional labour (Wilton 2008) in the form of requesting accommodation and, in the process, disclosing a disability or impairment in order to perform actual labour in the form of participation for your research interest and gain before participation is even established. To reduce the amount of necessary emotional labour, possible accommodation needs to be communicated proactively:

- If research requires participants to travel to a particular location, for example to attend a focus group or meet for an interview:
  - Are parking spaces reserved for disabled drivers?
  - Are accessible toilets nearby?
  - Are the premises themselves accessible, that is can meeting rooms be reached without having to navigate stairs?
  - Are the doorframes wide enough for a wheelchair to pass through?
  - Could the premises accommodate the presence of possible company such as an assistant or a carer, for example are there possible waiting rooms nearby?
- Could participants request to have a sign language interpreter be present for the interview or focus group? Are there induction loops?
- Could participants request any provided material (e.g., consent forms, participant information letter) in advance and in different formats, for example large print / in different colour schemes / in Braille?
- For interviews, could participants request the interview be conducted as a (audio-) computer-assisted self-interview?

If budget or other constraints do not allow for any or all of these accommodations, it is equally important to note this in any initial communication to reduce the participants' need to enquire and therefore disclose well in advance of the actual participation. Olkin (2004) observes that participants, especially in the deaf and hearing-impaired community, note the presence or absence of possible accommodations, and that their willingness to participate is strongly affected by how the researcher's efforts are perceived. Equally, participants note whether any concerted effort was made to reach out to the community and whether care was taken to word participant invitation and communication in a tactful and respectful manner (Olkin 2004). Suitable accommodations for hearing-impaired participants may include offering a communication channel to the researcher through voice relay services, such as TTY, textphone, teletypewriter or minicom (Farmer and Macleod 2011; Olkin 2004) or the option to request sign language interpreters (Olkin 2004).

The question whether to pay participants is a contentious one (see, e.g., Head 2009; Belfrage 2016, Bentley and Thacker 2004) though a distinction should be made between reimbursing expenses and paying for the participants' time. To make research more accessible, it seems unavoidable to offer full reimbursement for expenses. Not only does this help participants with sensory or mobility issues, for whom extra costs can add up quickly, for example, through the use of public transport or taxi or the need to be accompanied by a carer or an assistant; it also increases the chance of access for poorer and marginalised groups (see, e.g., Reid 1993, on the marginalisation of poor women in Psychology research). On that basis, it could be argued that payment should also be offered for the participant's time, effort and (emotional) labour and that not doing so would be exploitative (Farmer and Macleod 2011). However, the Turner and Beresford (2005) report from a consultation with participants stated that some felt 'service users may feel pressured to give more than they are able to give because of payments' (p. 42). This would be a particular concern for participants with health issues that made them prone to fatigue and exhaustion, who may feel obliged to continue to the detriment of their own wellbeing, and requires care and attention on the researcher's side.

Lastly, it is unfortunately also the case that some types of payment may be to the detriment of participants who are receiving disability or unemployment benefits, potentially putting an already vulnerable and marginalised group at further risk; it is therefore necessary to check with the individual participant how they can be offered adequate and fair compensation without putting them at legal or financial risk.

10.3.3   Scheduling for Interviews and Focus Groups

Allowing for more time throughout the entire research is a relatively straightforward way for you to make it easier for people to participate. If you schedule direct contact with participants, that is, for interviews (face-to-face, through Skype or by phone) or focus groups, and the time of contact is set by you rather than the participant, smaller groups and more frequent breaks will make it easier for all participants involved. In focus groups with elderly participants, Barrett and Kirk (2000) noticed that participants tended to lose concentration and tire fast and that with increasing group size, participants struggled to understand researchers and other participants; they consequently suggest to plan shorter sessions and refreshment breaks to allow their participants to recover. For a similar reason, Kroll (2011) recommends smaller group sizes to accommodate the individual participant. Although Barrett and Kirk's research was specific to elderly, disabled participants, shorter sessions and frequent breaks will also make it easier for participants with sensory impairments—as Farmer and Macleod (2011) point out, lip-reading or following a sign translator is tiring—or in fact for any participant who is prone to exhaustion, for whichever reason.

Being able to choose from a range of different times during the day over a range of different days benefits most participants: It allows participants with varying levels of energy to attend at a time they feel most comfortable with, but also accommodates different work patterns and lifestyles. To participate during working hours, potential participants will need to have the flexibility to take time off work and the financial means to buffer the potential loss of income from doing so; restricting the time available to working hours can therefore potentially disadvantage individuals with lower incomes or precarious employment situations. Awareness of school holidays, religious holidays or requirements to observe religious practices, for example Islamic prayer (Salah) times, can also increase the number and scope of potential participants. Even taking all this into consideration, flexibility and a certain amount of over-recruiting may be required. Barrett and Kirk point out that particularly where disabled and elderly participants and their carers are involved, cancellation due to illness (the participant's or their carer's) is likely.

Lastly, Kroll (2011) notes that it is not just the communication with participants and data collection that may take more time, but also data handling and analysis. Preparing information in different formats and modes; hiring and coordinating with a sign language translator (see below); and transcribing across multiple modes of participation all further extend the time needed to conduct a research project.

10.3.4   Choosing Accessible Premises

The premises and locations where research is conducted are not always under your control; for example, when conducting ethnographic research, the location is determined by the nature and scope of your research question. At other times, you might offer to meet participants at a location of their choice, such as their home or workplace; in this context, it will be more likely that they will have adjusted their environment to suit their accessibility needs. But in some instances, for example when conducting focus groups, it is inevitable to ask participants to attend a location that you have set. Kroll (2011) recommends that any such location fulfils a number of requirements:

- Easily accessible, especially by public transport
- Accessible parking nearby
- Accessible bathrooms in reasonable proximity to the room or location where the meeting takes place
- Few physical barriers for the participant to navigate
- Well lit to allow to create optimum conditions for participants with auditory or visual impairments

It may well be that some or all of these aspects are not under your control. Barrett and Kirk, (2000), for example further suggest to strive for an environment with little reverb: 'a quiet and echo-free environment is essential. Reverberation in a room can be reduced by the use of carpets, curtains, plants' (p. 627). While you may have some flexibility in choosing a room, you may not always have the added opportunity to adjust its furnishing. However, as I have emphasised above, the key is good advance communication with participants and to inform them as much as you can of any features you consider either particularly beneficial or detrimental. To keep with Barrett and Kirk's example, this might mean giving your participants advance notice that the location would be less suitable for participants with hearing impairments. Informed participants are in a better position to decide whether to attend or not.

10.3.5   Preparing Accessible Visual Material

This section concerns visual material which may be provided to participants throughout the research process, ranging from participant information and consent letters to surveys, research findings or a summary which may be made available to participants, or any material that may be presented to participants in order to elicit a comment or response. In this context, visual material includes both printed material and content presented on a screen.

To make printed material accessible, Farmer and Macleod (2011) and the UK Association for Accessible Formats (UKAAF 2012) recommend a clear, that is, uncluttered, layout with maximum readability. This can be achieved through a variety of measures:

- Font size 14 or up[1]
- Minimal use of emphasis such as italic, bold or underlined text
- Adequate spacing of letters and lines
- Use of clear, sans serif fonts as sans serif fonts are more legible for people with visual impairments and are beneficial for dyslexic participants
- Minimal use of graphics; where graphics are used, making sure that text is not broken up too much by graphics
- High contrast, low or no glare.

Additionally, the choice of colour and colour schemes needs to be carefully considered. Not only are some colours and colour schemes less legible than others (see, e.g., Gradišar et al. 2007; Greco et al. 2008) but where meaning is communicated through colour, for example, through the use of using red to highlight critical or wrong statements and green to highlight correct or acceptable, colour-blind participants will be less or not able at all to extract that meaning. Several free websites offer excellent resources to check colours and colour palettes for accessibility.

Should you offer a Braille version? Farmer and Macleod (2011, p. 47) suggest you do, while at the same time cautioning that only 'around 3 per cent of people registered blind or partially sighted use Braille'; offering

Braille is therefore only of limited help to a large number of partially sighted participants but will be of substantial assistance to those who are able to read it.

For many partially sighted or blind participants, accessibility can be improved by providing printed material in digital form instead, as this allows the use of screen readers, that is, assistive software which translates content on the screen into speech read out loud to the user. However, this approach comes with its own set of problems. Lazar et al. (2007, p. 247) report on the most frustrating issues for screen reader users:

> The top causes of frustration reported were (a) page layout causing confusing screen reader feedback; (b) conflict between screen reader and application; (c) poorly designed/unlabelled forms; (d) no alt text for pictures; and (e) 3-way tie between misleading links, inaccessible PDF, and a screen reader crash.

For online content, the World Wide Web Consortium (W3C) standard sets guidelines for accessible web content[2]; For electronic documents, the ICT for Information Accessibility in Learning project has collated a list of resources[3]; this includes providing the right kind of metadata for electronic documents, for example tags for pdf documents (Brady et al. 2015; Drümmer and Chang 2014; Bigham et al. 2016).

10.3.6 Accessible Verbal Communication and Auditory Material

The previous section primarily discussed material that was presented to participants in printed form or on a screen. The current section relates to the simple spoken word, whether by the individual researcher or in the context of a group discussion.

Hearing-impaired participants benefit from a suitable choice of premises, which means well lit (Kroll 2011; Farmer and Macleod 2011), little reverb (Barrett and Kirk 2000) and preferably the presence of induction loops (Farmer and Macleod 2011); in addition, the use of visual aids can further aid understanding (Farmer and Macleod 2011).

Although deaf and hard-of-hearing participants may use lip-reading to aid understanding, in addition to it being very tiring (Farmer and Macleod 2011), Whitney (2006) warns that it is unreliable as an exclusive source of understanding: 'As vowels do not appear on the lips, much of lip-reading is guess work, based on the context or situation. In perfect conditions (good lighting, no moustaches, etc), lip-reading is still 70% guesswork' (p. 289). For optimal conditions for lip-reading, the speaker, that is, the person whose lips are read, would also have to be trained to enunciate very clearly (Farmer and Macleod 2011).

Another option to improve accessibility is the provision of a sign language interpreter (Olkin 2004; Farmer and Macleod 2011) although this requires additional considerations:

- Like other languages, sign language exists in different forms and dialects, for example British Sign Language, American Sign Language and Australian Sign Language. When hiring a sign language interpreter, you therefore need to make sure to approach an interpreter with the relevant training; additionally, participants may have a preference for interpreters already known to them, or, conversely, be reluctant to be interpreted by people they already know (Olkin 2004).
- Some terms may not have a corresponding expression in the relevant sign language; something that would have to be discussed in advance between researcher and interpreter (Farmer and Macleod 2011).

In recent years, Computer-Assisted-Self-Interviewing (CASI) has increasingly been used (Couper et al. 2003; Leeuw et al. 2003) as a means of eliciting responses on sensitive topics: The interviewer can type in a question or set up the computer with a list of questions in advance, before handing over the computer for the participant to respond; in the Audio-CASI variant, the question is read out to the participant via headphone. Both variants are recommended by Farmer and Macleod (2011) as a means to make research methods more accessible: with AUDIO-CASI (or ACASI), appropriate technology can provide amplification for the audio output to assist hearing-impaired participants, whereas deaf participants can read the interviewer's questions from the screen; participants with speech impediments may prefer to enter their responses directly rather than in oral communication with the interviewer. Note, however, that self-entry requires a certain level of manual dexterity and fine-motor control and that some participants may require extra help with data entry, such as adapted keyboards or pointers, which may also take more time (Farmer and Macleod 2011).

Although participants with speech impediments potentially benefit from employing a CASI or Audio-CASI setup, it could be argued that this puts the onus on participants to make themselves understood. However, Farmer and Macleod (2011) point out that the process of establishing understanding between participant and researcher is likely exhausting for both parties, requiring more time and more breaks, and therefore recommend questions requiring short(er) answers. Ultimately, it has to be the participants' choice.

Barrett and Kirk (2000) further note that speech impediments may require further accommodation in the context of group discussions, and that time and effort to transcribe data will increase.

Lastly, Kroll (2011, p. 72) suggests that some light rephrasing of questions, for example to reduce the number of sibilants, can aid understanding for hearing-impaired participants:

Example (High frequency sounds): How satisfied are you with the overall quality of care you receive? Are you satisfied, somewhat satisfied, neither satisfied nor dissatisfied, or very dissatisfied?

Example (low frequency sounds): How would you rate the overall quality of the medical care you get? Is it excellent, very good, good, fair or poor?

Admittedly, this is feasible only in a context where much of the content of the uttered speech is known in advance, such as in a structured interview or survey, whereas a semi-structured or unstructured interview or a discussion in a focus group would not allow for the required amount of planning and careful tailoring of speech.

10.4   General Communication with Participants

Accessible research requires flexibility and attention to the needs of participants. This includes being attentive to participants' ability to participate without detriment to their health. Parsons et al. (2000) stress the need for the interviewer to be trained not only in meeting the participants' communication needs but also in spotting signs of exhaustion and fatigue. For example, Barrett and Kirk (2000) observed elderly participants losing concentration and focus towards the end of the focus group sessions, leading to participants getting distracted and losing their line of thought, particularly when being interrupted by the interviewer or other participants. In response, Barrett and Kirk stressed the importance of clear, simple and understandable questions and made sure to introduce changes of topics and provide plenty of signposting.

Related, Gregory (1996, as cited in Farmer and Macleod 2011) recommends a range of rules to make information more accessible:

- Don't use long sentences
- Include one main point, and only one or two clauses in a sentence
- Communicate in the active voice, rather than the passive
- Avoid abstract concepts
- Use simple words, without being patronising
- Repeat difficult or unfamiliar words
- Don't use jargon
- Avoid abbreviations and acronyms
- Avoid using the third person

Although those rules were specifically devised for communicating with people with learning disabilities, arguably most of them are applicable and helpful for communication in a wide variety of contexts[4] though not always feasible, particularly in the context of spontaneous and unplanned speech.

10.5   Presenting and Disseminating Research

The final section of this chapter outlines recommendations to improve research accessibility by improving the accessibility of research findings. Wilson et al. (2013) argued that '[people with disabilities] are often excluded from participation in political dialogues and democratic processes, and hence decision making processes that greatly affect them'. By the same token, where research findings are inaccessible, people are excluded from the public discussion of these findings and, in turn, are not given all the relevant information to make decisions.

Several recommendations on how to improve the accessibility of written material have already been covered in a previous section. Undoubtedly, sometimes, these will be outside your control, as most journals impose strict formatting guidelines for paper submissions and very few journals consider accessibility part of their submission guidelines (e.g., Garbutt 2009). But research dissemination means more than papers in peer-reviewed journals: It also means disseminating findings through information leaflets for participants; presentations and hand-outs in workshops—and storing information on departmental homepages or on websites such as academia.edu or researchgate.net. In part, dissemination *is* accessibility.

And yet, even the most commonly used format (Portable Document Format or PDF) is fraught with problems (Bigham et al. 2016). Bigham et al. (2016, p. 622) list a range of requirements for accessible PDF documents:

> [...] visual information should be represented in another form (often, plain text) so that it can be consumed in another way by blind or visually impaired readers. [...] The document should allow for easy magnification for people with low vision. While one may easily mistake this for a zoom feature in a PDF reader, which the current format does not due to (among other things) its two column format. To be made easier to read by people with reading disabilities, the document should be flexible, allowing for not only magnification but changing of fonts, colors, and layout.

Although good guidance is available to making PDF documents accessible (e.g., Drümmer and Chang 2014), Bigham et al. note that this is only a partial solution since Cascading Style Sheets (CSS) or Hypertext Markup Language (HTML) are superior to PDF and conclude that '[t]he most frustrating part about the lack of accessibility of research papers is that all we would need to do to make them accessible is to decide to do so' (Bigham et al. 2016, p. 628).

10.6   Conclusion

Bigham et al.'s quote in the last paragraph is pertinent to the argument I have been trying to set out within this chapter. First, accessible research is *necessary*. If you are serious about your research representing and examining the lived experience of your participants, making research accessible and communicating possible adaptations and accommodations is your responsibility. Second, accessible research is *possible*. It does require planning and attention; it may require extra funding, and it almost certainly takes more time, but it is feasible. Third, making research more accessible is *productive*. Lowering barriers to research participation is not a zero-sum game. Flexible schedules can improve access for people with chronic illnesses as well as for people with caring responsibilities or those who cannot afford to take time off work; while shorter focus group sessions with more breaks also benefit elderly participants or people who are pregnant.[5] Enabling more people to participate in your research will help you make more voices heard. These voices have always been there; it is time we start to listen.

Notes

1. Of course, this can always be subject to publication layout limitations, as is the case with the chapter in this book.
2. A good overview is listed here: https://www.w3.org/standards/webdesign/accessibility.
3. List of resources here: http://www.ict4ial.eu/guidelines/making-electronic-documents-accessible/resources-help-make-electronic-documents-accessible.

4. In addition, reducing the extent of the use of metaphors or humour could benefit non-neurotypical participants (Samson and Hegenloh 2010; Lyons and Fitzgerald 2004).
5. I use 'people who are pregnant' rather than 'pregnant women' to acknowledge pregnant trans men.

References

Abberley, P. (1992). Counting Us Out: A Discussion of the OPCS Disability Surveys. *Disability, Handicap & Society, 7*(2), 139–155.

Barrett, J., & Kirk, S. (2000). Running Focus Groups with Elderly and Disabled Elderly Participants. *Applied Ergonomics, 31*(6), 621–629.

Belfrage, S. (2016). Exploitative, Irresistible, and Coercive Offers: Why Research Participants Should Be Paid Well or Not at All. *Journal of Global Ethics, 12*(1), 69–86.

Bentley, J. P., & Thacker, P. G. (2004). The Influence of Risk and Monetary Payment on the Research Participation Decision Making Process. *Journal of Medical Ethics, 30*(3), 293–298.

Bigby, C., Frawley, P., & Ramcharan, P. (2014). Conceptualizing Inclusive Research with People with Intellectual Disability. *Journal of Applied Research in Intellectual Disabilities, 27*(1), 3–12.

Bigham, J. P., Brady, E. L., Gleason, C., Guo, A., & Shamma, D. A. (2016). *An Uninteresting Tour Through Why Our Research Papers Aren't Accessible.* Proceedings of the 2016 CHI Conference Extended Abstracts on Human Factors in Computing Systems – CHI EA '16, San Jose, 621–631.

Brady, E., Zhong, Y., & Bigham, J. P. (2015). *Creating Accessible PDFs for Conference Proceedings.* Proceedings of the 12th Web for All Conference on – W4A '15, Florence, 1–4.

British Academy of Management. (2013). *The British Academy of Management's Code of Ethics and Best Practice.* London: BAM.

British Sociological Association. (2006). *Statement of Ethical Practice for the British Sociological Association–Visual Sociology Group.* Durham: BSA.

Cheyns, E. (2014). Making 'Minority Voices' Heard in Transnational Roundtables: The Role of Local NGOs in Reintroducing Justice and Attachments. *Agriculture and Human Values, 31*(3), 439–453.

Couper, M. P., Singer, E., & Tourangeau, R. (2003). Understanding the Effects of Audio-CASI on Self-Reports of Sensitive Behavior. *Public Opinion Quarterly, 67*(3), 385–395.

Drümmer, O., & Chang, B. (2014). *PDF/UA in a Nutshell: Accessible Documents with PDF* https://www.pdfa.org/wp-content/until2016_uploads/2013/08/PDFUA-in-a-Nutshell-PDFUA.pdf. Last accessed March 31, 2017.

Easton, C. (2013). An Examination of the Internet's Development as a Disabling Environment in the Context of the Social Model of Disability and Anti-Discrimination Legislation in the UK and USA. *Universal Access in the Information Society, 12*(1), 105–114.

Economic and Social Research Council. (2010). *Guidance Note for Researchers and Evaluators of Social Sciences and Humanities Research.* http://ec.europa.eu/research/participants/data/ref/fp7/89867/Social-sciences-humanities_en.pdf. Last accessed 13 Sept 2017.

Farmer, M., & Macleod, F. (2011). *Involving Disabled People in Social Research.* Office for Disability Issues.

Garbutt, R. (2009). Is There a Place within Academic Journals for Articles Presented in an Accessible Format? *Disability & Society, 24*(3), 357–371.

Gradišar, M., Humar, I., & Turk, T. (2007). *The Legibility of Colored Web Page Texts.* Proceedings of the International Conference on Information Technology Interfaces, ITI, Cavtat, 233–238.

Greco, M., Stucchi, N., Zavagno, D., & Marino, B. (2008). On the Portability of Computer-Generated Presentations: The Effect of Text-Background Color Combinations on Text Legibility. *Human Factors, 50*(5), 821–833.

Hammel, J., Magasi, S., Heinemann, A., Gray, D. B., Stark, S., Kisala, P., Carlozzi, N. E., Tulsky, D., Garcia, S. F., & Hahn, E. E. (2015). Environmental Barriers and Supports to Everyday Participation: A Qualitative Insider Perspective From People with Disabilities. *Archives of Physical Medicine and Rehabilitation, 96*, 578–588.

Head, E. (2009). The Ethics and Implications of Paying Participants in Qualitative Research. *International Journal of Social Research Methodology, 12*(4), 335–344.

Jones, M. K. (2016). Disability and Perceptions of Work and Management. *British Journal of Industrial Relations, 54*(1), 83–113.

Kavanagh, A. M., Krnjacki, L., Aitken, Z., Lamontagne, A. D., Beer, A., Baker, E., & Bentley, R. (2015). Intersections between Disability, Type of Impairment, Gender and Socio-Economic Disadvantage in a Nationally Representative Sample of 33,101 Working-Aged Australians. *Disability and Health Journal, 8*(2), 191–199. Elsevier Inc.

Kroll, T. (2011). Designing Mixed Methods Studies in Health-Related Research with People with Disabilities. *International Journal of Multiple Research Approaches, 5*(1), 64–75.

Lazar, J., Allen, A., Kleinman, J., & Malarkey, C. (2007). What Frustrates Screen Reader Users on the Web: A Study of 100 Blind Users. *International Journal of Human-Computer Interaction, 22*(3), 247–269.

Leeuw, Edith de, Hox, J., & Kef, S. (2003). Computer-Assisted Self-Interviewing Tailored for Special Populations and Topics. *Field Methods, 15*(3), 223–251.

Lockyer, S. (2015). 'It's Really Scared of Disability': Disabled Comedians' Perspectives of the British Television Comedy Industry. *The Journal of Popular Television, 3*(2), 179–193.

Lyons, V., & Fitzgerald, M. (2004). Humor in Autism and Asperger Syndrome. *Journal of Autism and Developmental Disorders, 34*(5), 521–531.

McDonald, K. E., & Keys, C. B. (2008). How the Powerful Decide: Access to Research Participation by Those at the Margins. *American Journal of Community Psychology, 42*(1–2), 79–93.

McDonald, K. E., Keys, C. B., & Balcazar, F. E. (2007). Disability, Race/Ethnicity and Gender: Themes of Cultural Oppression, Acts of Individual Resistance. *American Journal of Community Psychology, 39*(1–2), 145–161.

Mik-Meyer, N. (2015). Gender and Disability: Feminizing Male Employees with Visible Impairments in Danish Work Organizations. *Gender, Work and Organization, 22*(6), 579–595.

Mik-Meyer, N. (2016). Disability and 'Care: Managers, Employees and Colleagues with Impairments Negotiating the Social Order of Disability. *Work, Employment & Society, 30*(6), 1–16.

National Union of British Sign Language Interpreters. (2017). *Freelance Fees for Interpreting Engagements for BSL/English Interpreters.* Accessed March 28. http://www.nubsli.com/guidance/interpreter-fees/.

Nind, M. (2008). Learning Difficulties and Social Class: Exploring the Intersection through Family Narratives. *International Studies in Sociology of Education, 18*(2), 87–98.

Nind, M., & Vinha, H. (2012). *Doing Research Well? Report of the Study: Quality and Capacity in Inclusive Research with People with Learning Disabilities.* https://www.southampton.ac.uk/assets/imported/transforms/contentblock/UsefulDownloads_Download/97706C004C4F4E68A8B54DB90EE0977D/full_report_doing_research.pdf Last accessed March 30, 2017.

Nind, M., & Vinha, H. (2013). Methodological Review Paper. Practical Considerations in Doing Research Inclusively and Doing It Well: Lessons for Inclusive Researchers. National Centre for Research Methods: Methodological Review Paper. http://eprints.ncrm.ac.uk/3187/1/Nind_practical_consider-ations_in_doing_research_inclusively.pdf. Last accessed 30 Mar 2017.

Office for National Statistics. (2014). *Official Statistics: Disability Facts and Figures.* https://www.gov.uk/government/publications/disability-facts-and-figures/disability-facts-and-figures. Last accessed September 13, 2017.

Olkin, R. (2004). Making Research Accessible to Participants with Disabilities. *Journal of Multicultural Counseling and Development, 32,* 332–343.

Parsons, J. A., Baum, S., & Johnson, T. P. (2000). *Inclusion of Disabled Populations in Social Surveys: Reviews and Recommendations.* Chicago: Survey Research Laboratory, University of Illinois for the National Center for Health Statistics.

Petersen, A. J. (2012). Imagining the Possibilities: Qualitative Inquiry at the Intersections of Race, Gender, Disability, and Class. *International Journal of Qualitative Studies in Education, 25*(6), 801–818.

Postle, K., Wright, P., & Beresford, P. (2005). Older People's Participation in Political Activity—making Their Voices Heard: A Potential Support Role for Welfare Professionals in Countering Ageism and Social Exclusion. *Practice, 17*(3), 173–189.

Reid, P. T. (1993). Poor Women in Psychological Research: Shut Up and Shut Out. *Psychology of Women Quarterly, 17*(2), 133–150.

Samson, A. C., & Hegenloh, M. (2010). Stimulus Characteristics Affect Humor Processing in Individuals with Asperger Syndrome. *Journal of Autism and Developmental Disorders, 40*(4), 438–447.

Schroedel, J. G. (1984). Analyzing Surveys on Deaf Adults: Implications for Survey Research on Persons with Disabilities. *Social Science and Medicine, 19*(6), 619–627.

Turner, M., & Beresford, P. (2005). *User Controlled Research: Its Meanings and Potential.* Commissioned Report for INVOLVE.

UKAAF. (2012). *Creating Clear Print and Large Print Documents.* http://www.ukaaf.org/wp-content/uploads/2014/09/G003-UKAAF-Creating-clear-print-and-large-print-documents.pdf. Last Accessed 30 Mar 2017.

Ville, I., & Ravaud, J. F. (1998). Work Values: A Comparison of Non-Disabled Persons with Persons with Paraplegia. *Disability and Rehabilitation, 20*(4), 127–137.

Wellcome Trust. (2017). *Guidelines on Good Research Practice.* https://wellcome.ac.uk/funding/managinggrant/guidelines-good-research-practice. Last accessed March 30, 2017.

Whitney, G. (2006). Enabling People with Sensory Impairments to Participate Effectively in Research. *Universal Access in the Information Society, 5*(3), 287–291.

Wilson, E., Campain, R., Moore, M., Hagiliassis, N., McGillivray, J., Gottliebson, D., Bink, M., Caldwell, M., Cummins, B., & Graffam, J. (2013). An Accessible Survey Method: Increasing the Participation of People with a Disability in Large Sample Social Research. *Telecommunications Journal of Australia, 63*(2), 24.1–24.13.

Wilton, R. D. (2008). Workers with Disabilities and the Challenges of Emotional Labour. *Disability & Society, 23*(February 2015), 361–373.

11

Ethics in Qualitative Research

Sylwia Ciuk and Dominika Latusek

11.1 Introduction

Qualitative researchers working in the diverse field of social sciences need to address ethical issues at every stage of the research process (Clegg and Slife 2009), regardless of the perspective, research design or methods of data collection they opt for. As is widely recognised, ethical thinking in qualitative research goes beyond ethical decisions during data collection and analysis (Kara and Pickering 2017). Indeed, it concerns broader issues, such as presentation and dissemination of research results, public engagement or the deposition of data in research databanks in order to make them available for other researchers, which is increasingly required by funding bodies. In the light of the rapidly changing research landscape that has, in many contexts, become subject to stringent formal ethical review and governance and where technological advances have offered

S. Ciuk (✉)
Oxford Brookes University, Oxford, UK

D. Latusek
Kozminski University, Warsaw, Poland

© The Author(s) 2018
M. Ciesielska, D. Jemielniak (eds.), *Qualitative Methodologies in Organization Studies*,
https://doi.org/10.1007/978-3-319-65217-7_11

new possibilities for research innovation, long-standing ethical issues have taken on new meanings and new ethical dilemmas have emerged (Mauthner et al. 2012). For example, visual methods, many of which are innovative and pioneering (Howell et al. 2014), often force researchers to reconsider their responses to a range of crucial ethical issues, such as informed consent, confidentiality or ownership, along with questions around data presentation and dissemination (e.g. Allen 2012; Cox et al. 2014). Similarly, academics conducting internet research, the context and scope of which have grown exponentially, need to navigate complex ethical terrains. In this context, Birch et al. (2012, p. 4) might indeed be right when they observe that 'ethics matter more now than they did a decade ago'.

The significance of ethical issues in empirical research, employing both qualitative and quantitative methodologies, is reflected in the plethora of codes of ethics put forward by relevant professional associations (e.g. Association of Internet Researchers and British Sociological Association or its American counterpart) and the growing institutional regulations which increasingly not only seek to guide but more recently also to govern the work of researchers. Although the importance of ethical guidance and the associated relevant training for researchers is rarely, if ever, disputed, the institutional approach to research ethics, sometimes referred to as ethics (Haggerty 2004) and 'audit creep' (Stanley and Wise 2010, p. 25), has come under strong criticism (for a more detailed discussion, see e.g. Cannella and Lincoln 2007; Coupal 2005; Hammersley 2009; Hedgecoe 2008). As authors point out (e.g. Birch et al. 2012; Hammersley and Trainou 2012; Stanley and Wise 2010), the idiosyncratic qualities of qualitative research which typically involve a considerable degree of flexibility of the research design, and the collection of (more or less) unstructured data that is typically collected in natural settings are not easily reconciled with the standardised and largely inflexible external formal ethical regulation that relies on universalist principles and generalist criteria. This recognition notwithstanding, the remit of Research Ethics Committees (RECs) and the associated anticipatory, pre-study ethics regulation (Mautghner et al. 2012) has been expanding. This trend, however, has not reduced the ethical challenges faced by qualitative researchers and there is still considerable ambiguity surrounding ethical decision-making as more 'than one set of norms, values, principles and usual practices can be seen to legitimately apply to the issue(s) involved' (Markham and Buchanan 2012, p. 5). In fact, as some authors observe (e.g. Tilley and Woodthorpe 2011), ethical governance can at times exacerbate, rather than reduce ethical tensions and can itself pose new ethical dilemmas. For example, researchers can be required to deposit their data in research databanks so that others could reuse them in the future. This requirement, however, as Mauthner et al. (2012, p. 180) observe, 'raises ethical and moral issues about the responsibility that we take as researchers for the methods we use; for how we carry our research and for the context in which this occurs'. Ethical requirements associated with procedural ethics (Guillemin and Gillam 2004) therefore need to be supplemented with professional guidance, theoretical models and 'contextualized reasoning' (Birch et al. 2012, p. 6) which can help researchers negotiate ethics in practice (Guillemin and Gillam 2004) in dynamic research settings (see also Markham and Buchanan 2012).

In the remainder of this text, we revisit some of the most commonly recurring ethical issues facing qualitative researchers at different stages of the research process and point to some of the new ethical dilemmas associated with the changing research landscape and innovative research methodologies. Those interested in the philosophical considerations regarding research ethics may refer to the more specialist sources, such as Kent (2000), Christians (2011) or Hammersley and Traunou (2012). Similarly, others looking for more detailed discussions of ethical challenges linked to specific methodologies and approaches can consider the already available relevant sources, such as Cox et al. (2014) for guidelines for ethical visual research methods or Markham and Buchanan (2012) for recommendations for internet research. The questions we focus on are intended to serve as an illustration of the multitude of ethical issues that social scientists are faced with in their daily practice, rather than a comprehensive review.

11.2   Informed Consent

One of the most central requirements of research ethics committees, and arguably also one of the most often debated ethical challenges, is the need to obtain informed consent from the prospective research participants.

Despite the seemingly commonsensical nature of this postulate according to which those invited to take part in research have to give their (usually written) consent to participate in the project after having been informed about the nature, purpose and outcome of the study and their role in it, as well as the possibility to withdraw from the study (and the unprocessed data that they helped generate), its implementation in different research settings is certainly far from straightforward, in particular when considered in the context of longitudinal research projects, or with studies utilising more participative methodologies.

While ethics committees typically require obtaining consent once, usually prior to data collection, compelling arguments have been put forward suggesting that consent should rather be regarded as a matter of ongoing negotiations between the researcher(s) and the research participants and revisited as the research evolves. According to Birch and Miller (2002), for example, a one-off consent—regardless of whether given orally or made in writing—is insufficient, especially when it comes to ethnographic studies or other kinds of longitudinal qualitative research projects which require longer-term involvement of the research participants. Participation in such studies requires a different kind of commitment and that is why, as argued by Birch and Miller (2002), it should be subject to continuous renegotiation. Furthermore, researchers have an obligation to remind their participants on a regular basis that they may revoke their consent at any time. This point is further reiterated by Neale (2013, p. 6) who observes that 'well established ethical principles [such as consent] take on new meaning and need reworking when seen with a temporal gaze'. In longitudinal projects, the ethical landscape gets broadened and becomes further complicated. Participants may choose to withdraw temporarily from the research project but simultaneously reserve the right to re-join at a later stage. When the time frames of research get extended, so does the likelihood that earlier unforeseen ethical challenges will emerge (op cit). But it is, however, not only the consent to participate in a research project that may be subject to renegotiation between researchers and their participants. As a research project evolves, the nature and the scope of participation may also need to be renegotiated. As already mentioned, many researchers point out (e.g. Miller and Bell 2002; Duncombe and Jessop 2002) that at the beginning of a research project both researchers and research subjects are not always able to accurately assess the potential impact of their research on their participants, nor are the participants often in a position to fully grasp what taking part in the study entails. It is also important to remember that research participants' personal circumstances may also change with time, which may, in turn, have an impact on their participation.

In the context of visual research methods, Cox et al. (2014, p. 12) propose to view consent as 'a series of decision that take place at pre-identified points as project unfolds'. As the authors outline, in visual methods, consent refers not only to the generation and collection of visual images, but, importantly, also applies to their analysis, presentation and crucial dissemination among different audiences. Cox et al. (2014) therefore recommend seeing consent as composing of different levels and stages. This point is well illustrated by Murray and Nash's (2016) paper discussing the ethical challenges of photovoice and photo elicitation in two separate studies carried out by the quoted authors, one of which explored the embodiment of pregnancy in Australia, whereas the other focused on the experiences of infant setting in Vietnam. The authors explain how in the Australian study, three different consent forms were used at different stages of the research process, which not only sought to explain participants' rights and responsibilities (consent one), but also focused on obtaining consent from others who appeared in the photographs (consent two). In particular, the last stage of negotiating consent described by the authors is instructive. Murray and Nash (2016) describe a detailed process of consulting with participants the extent to which they consented to their different images being disseminated and shared with the academic audience and the general public, a procedure of 'different levels of consent' also usefully described by Lunney et al. (2015) in the case of photo elicitation research into young women's experiences with drinking alcohol. Another challenge related to the requirement of obtaining informed consent from research participants is linked to wider cultural and institutional norms of a given research setting, which, in some situations, might run counter to this requirement. For example, Marzano (2007) conducted a research project into the experiences of the terminally ill at an oncological ward in Italy. However, in Italy, at the time of data collection, the dominant institutional norm, as Marzano explains, was not to inform terminally ill patients of their actual condition. In such institutional contexts where the commonly held belief about the detrimental effects of revealing to patients information about their condition is deeply rooted in the local culture, the researcher may decide not to violate this norm and, therefore, not to disclose the purpose of their research. Marzano admitted that if he had revealed the real purpose of his research, he might have been forced to leave the research site and, most likely, might also have been forbidden to access it again so he decided to observe the cultural and institutional norm to withhold crucial information from the studied patients. As the author revealed, his decision came at a considerable emotional cost to him. The requirement to obtain informed consent from participants—like many other ethical issues—becomes further complicated in the context of internet-mediated research projects. As argued, among others, by Ellen Whiteman (2007), when doing (non-participant) observation online, it is not always clear what can be considered a private and what a public domain. Is it necessary to reveal one's identity when researching internet forums or online community networks? Or can we recognise them as public domains that may be monitored without the need to inform our research participants of the conducted study? There are different recommendations on this matter. Kozinets (2002, 2015), for example, writing about netnography and netnographers as cultural participants rather than unobtrusive observes calls for a full disclosure of researcher's identity and explains that even though information posted online is generally freely available, this fact does not mean that its authors automatically consent to it being used for academic purposes. According to Kozinets (2015, p. 139), the internet 'is not either public or private; it does not simply contain data but digital doubles of our identities and selves' and therefore it requires more 'creative and bricolage-based solutions' to ethical dilemmas (op cit, p. 139). He further lists a range of strategies researchers can use to inform research participants of the researcher's identity (such as posting relevant information in status updates or next to the researcher's name or using pop-ups). Others (e.g. Langer and Beckman 2005) see the recommendations put forward by Kozinets (2002) as 'far too rigorous' (Langer and Beckman 2005, p. 195) and argue that netnography 'enables the researcher in an unobtrusive and covert way to gain deeper insights' into people's options and motives, whereas Roberts suggests that while deciding on whether to reveal their identity researches need to consider not only the assessibility of the online community to outsiders, but also the perceptions and perferences of its members, the permanance of data as well as the sensitivity of the topic at hand. Finally, and relatedly, the requirement to obtain 'informed consent' poses considerable challenges to conducting disguised observation. Even though disguised observation is increasingly seen as problematic, there are a number of excellent ethnographic studies developed on the basis of this method. Ethics committees usually underline the need to inform research subjects of the fact that there is a research project conducted with their involvement and of the purpose of any such research project. There are, however, exceptions to this rule. It might be possible to opt for disguised observation if there is no other suitable method of studying a given research problem. Disguised observation may be also performed in studies of public behaviour, as already indicated, where the identity of research participants will remain unknown. In such cases, the recommendation is to monitor the behaviour of research participants and to treat even potential signs of reluctance as a refusal to take part in the study. However, we can still wonder to what extent researchers can really trust their ability to accurately interpret the intentions of their research subjects. Is there really a guarantee that the identity of the research subjects will remain anonymous? These and many other questions show the complexity behind ethical considerations while doing field work—even when one tries to act according to the already strict guidelines.

Tina Miller and Linda Bell (2002) raise an important question of the role of 'gatekeepers' in obtaining participants' consent. One of them—Linda Bell—conducted an interview-based study of a group of Bangladeshi women living in southern England, the access to whom was secured by a person closely involved with the community of interest. However, although the gatekeeper's support made it possible for Bell to obtain access and to secure consent from the research participants, she quickly realised that participants were in fact rather reluctant to take part in her study. She came to a conclusion that their consent was largely motivated by respect and a sense of obligation towards the gatekeeper, rather than their actual willingness to take part in the project. The gatekeeper was held in high regard in the studied community because of her origin and background, and the related social status. Questions, therefore, arise as to the extent to which one can treat the obtained consent as a sign of participants' readiness to take part in the study of their own free will. Similar questions emerge in the context of research in organisational settings. To what extent, for example, do employees who have agreed to take part in a study upon a request set out by their superiors have an actual option to decline this invitation? How can we know whether their consent is not primarily driven by fear of being punished if they do refuse to participate?

11.3   Protection of Research Participants' Identity

It is widely accepted that researchers are obliged to protect their research participants (and themselves) against any undesirable effects of their study. The requirement of doing no harm and the obligation to protect one's research participants have contributed to the practice of treating the identity of research subjects (i.e. people, organisations and selected social groups) as confidential and substituting it with pseudonyms in reports and publications of the results. Granting research participants' anonymity often involves omitting or obscuring certain information in the publication of research findings that could make it possible to identify the said participants. However, despite the fact that anonymity has started to be perceived as an ethical norm, in the changing research landscape securing anonymity has been increasingly challenging. Internet research is a good example of the potency of this problem.

The longevity and ease of traceability of information published online coupled with the common requirement to aim to widely disseminate one's research results to a range of audiences mean that standard solutions of securing anonymity are often no longer fit for purpose and researchers can no longer control how the data they share is consumed and reproduced by others (e.g. Tilley and Woodthorpe 2011), including the research participants themselves (see e.g. Lunnay et al. 2015). While some practical solutions have been proposed on how to overcome these challenges, such as rephrasing or avoiding direct quotes (see e.g. Kozinets 2015), it has also been suggested that 'the standard of anonymity in the context of the twenty first century academic work may need to be rethought' (op cit, p. 1), as even seemingly anonymised data can contain sufficient information to lead to identification of a given participant (Markham and Buchanan 2012).

Similarly, the requirement to protect the identity of our research participants is also problematic in the context of participatory and emancipatory research methods where the issues of anonymity need to be evaluated against the notions of shared authorship and empowerment. As observed by Tilley and Woodthorpe (2011), in some contexts, the otherwise seemingly uncontentious principle of anonymity can be at odds with the aim of the research and the dissemination plan. Participatory research designs therefore typically seek to give participants' choice as to whether they wish to retain anonymity or whether they would rather their identity was disclosed (Tilley and Woodthorpe 2011). As Christians (2011, p. 66) observes, anonymising procedures which 'researchers consider innocent [can be] perceived by participants as misleading or even betrayal', especially when they see themselves as important contributors to the research project. In a similar vein, strategies of anonymising images in visual methodologies, such as blurring and obscuring techniques, do not also always offer full anonymity and can have the unintended consequences of compromising the authenticity of the image and dehumanising the research participants (Cox et al. 2014).

These challenges might best be taken into consideration already at the planning stage of the research and revisited when negotiating consent. Ethical dilemmas might also arise when the participants' right to anonymity is in conflict with the rights of other parties. One such situation is described by John Van Maanen (1983) in relation to his ethnographic study of the New York Police Department (NYPD). Van Maanen was asked to testify in the case of a man who had been battered by the police, an incident which Van Maanen witnessed. In this case, Van Maanen refused to cooperate with the police, in order to protect his research participant. His decision, however, can be interpreted as potentially detrimental to the broader community. Relatedly, Sabir and Sabir Ben-Yehoshua (2017) have illustrated how participants might deliberately seek to potentially compromise their own anonymity in order to punish a member of their family (e.g. an ex-husband). Should we then respect the wish of our research participants, or should we rather choose to protect them and their environment against their recommendation? How to act if the findings of our research may benefit the majority of the community under study, but harm its minority? The existing source literature does not offer straightforward answers to the above questions. Instead, ethical choices are seen as being highly context dependent and 'requiring contextualized reasoning' rather than an application of 'abstract rules and principles' (Birch et al. 2012, p. 6), as well as an informed dialogue between the research and their participants (Sabir and Sabir Ben-Yehoshua 2017).

11.4   Maintaining Relationships with Research Participants

Many ethical issues in qualitative research based on direct relationships with research participants are related to commencing, maintaining and ending those relations. Jean Duncombe and Julie Jessop (2002) note that ethical issues already emerge at the stage of preparation for fieldwork. It is commonly believed that the ability to establish relations with research participants is very important for qualitative researchers. The skill is often considered to be a prerequisite for building trust with participants which, in turn, is expected to help the researcher obtaining more honest answers from research participants and richer data. But such an instrumental approach to building and maintaining relationships with research participants raises some considerable ethical questions. Critics of this approach (e.g. the earlier cited Duncombe and Jessop 2002) point out that in this perspective, relationships with research participants are treated as a form of emotional labour (Hochschild 1983). In practice, 'establishing relationships' in the field may at times manifest itself as 'faking friendship' in the field and, as argued by its critics (e.g. Fine 1994), it may, in fact, be more common among field researchers than research reports would lead one to believe. Others, for example Beech et al. (2009), call on researchers to get more involved with the problems of the communities they research. They point to a possibility of establishing mutually beneficial relationships which not only help the researchers secure better access to data, but support the studied communities in solving local issues.

The complexity of relationships with research participants and the related ethical issues are also covered by Duncombe and Jessop (2002) who give an account of how their ability to 'establish relationships with research participants' enabled them to obtain much more information than their research participants were initially willing to share with them. However, the researchers also point to the negative consequences of this. Jessop, for instance, conducted an in-depth interview with a man who was left by his wife after a long marriage, and obtained an extensive account of his past experience that he had not even shared with his wife. The interview ended with the man bursting into tears and the researcher leaving with a sense of guilt. By quoting the above example, the authors contribute to calls for respecting the research participants' right to ignore their deepest thoughts if they wish to do so. They argue that no research should force participants into reflexivity they find unwelcome.

The matter of emotions in research is yet another ethical issue related to establishing relationships with research participants. This concerns both researchers' and research participants' emotions. According to guidelines included in various codes of ethics, research should not cause emotional harm to its participants. But researchers are not always able to foresee which of the questions might evoke a strong emotional reaction in the participants. For example, one of the authors of this chapter conducted a study into a culture of two different organisations. When she asked one of the research participants about her views on and experiences with the company value of 'care', the research participant unexpectedly burst into tears midway through her answer. While the researcher did not intend to raise any topics that could evoke a strong emotional response which did not even seem necessary in the study, it turned out that the above-mentioned and seemingly neutral question about the core values of the company reminded the research participant of her child's death—and of the support her superior at the time gave her child beforehand. To her, the superior's attitude was a real-life embodiment of the value of 'care'. The researcher's question led to an emotional tension and inadvertently made the research participant recall painful memories. So, what to do in such a situation? What could be considered ethical behaviour in this case? Stopping the interview? Turning the voice recorder off? Showing empathy, compassion? Proceeding with the subject? Maybe it would have been reasonable to continue probing the participant about her superior to thus allow the participant experience more positive emotions and help her compose herself. The researcher, surprised with the turn of events, let the research participant finish the topic, and went on to continue the interview, carefully monitoring the participant.

Two other researchers, Wendy Mitchell and Annie Irvine (2008), who also encountered strong emotional reactions among their research participants in the course of their studies, formulated similar ethical questions. Each of them reacted in a different way to the emotionally charged situations, which has prompted them to argue for the need for a more conscious approach to emotions management in research. They recommend trying to predict, as much as it is possible, the emotional responses of one's participants and considering how to react to them prior to data collection. It is important to remember that research participants might themselves be well placed to communicate to the researcher how they would prefer to proceed after an emotional encounter. Researchers identifying themselves with the feminist perspective draw attention to further ethical considerations when doing fieldwork. They not only endorse the basic principle of doing no harm to research participants, but they also call for the need to approach research participants with care. How exactly to use the principle of care in practice tends to be viewed differently by authors and can itself pose a number of ethical challenges (cf. Mitchell and Irvine 2008). If a research participant, for example, reveals, in the course of the interview, something that troubles them, should we offer our help if we are in a position to provide it? To what extent are we—as social researchers—in a position to offer emotional support to our research participants? How can we judge whether our research participants would actually welcome our offer to help? What possible consequences can researchers' involvement have?

The issue of relationships with research participants becomes even more pronounced in the case of longitudinal research projects or when researchers decide to extend the timeframe (and scope) of past studies in previously unplanned ways and, as a consequence, seek to trace past participants, the practicalities of which can, as Miller (2015: 2) admits, 'at times feel analogous to stalking'. It is important for researchers conducting extended field work to be able to set and manage the boundaries of relationships and mutual expectations (see e.g. Lunnay et al. 2015; Neale 2013). This need can appear quite early on in one's research project and one's career. One of us, for example, was faced with a challenging situation while conducting the third interview in her academic career. The researcher was to conduct a series of interviews with employees working in the same department in a large multinational corporation. As it turned out later, the department was also a place where three female employees battled fiercely for a managerial position. The first interviewee (let us call her Anna) offered the researcher a very warm welcome; she suggested they call each other by their first names, contrary to the local custom, and offered to allocate more time for the interview after its allotted time had run out. The interview was to continue after work in a nearby café. The inexperienced researcher was glad to take the opportunity and was happy with the friendliness she was approached with. The following day, however, when she met another interviewee, she realised her enthusiasm was premature. By that time (the morning of the following day), it appeared that almost all employees of the company were convinced that the researcher was 'Anna's friend', which triggered mistrust towards the researcher and made it virtually impossible to proceed with the research project.

11.5   Presentation of Research Findings

As has been stated earlier, guidelines on the dissemination and presentation of findings can at times run counter to the wishes of research participants. In particular, in participative and emancipatory research methods, such as photovoice, such discrepancies can pose considerable ethical dilemmas for the concerned researchers, as well captured by the earlier cited study by Murray and Nash (2016). The authors illustrate how the guidelines put forward by the British Sociological Association (2006) which, quite uncontroversially, outline that researchers must avoid potentially inappropriate or sexually explicit images, were seen as problematic in one of their studies (carried out by Nash) on pregnancy embodiment. As it turned out, a number of participants produced and shared nude images of themselves in order to fully address the question they were set to explore and to depict their lived embodied experiences of pregnancy. Contrary to the ethical guidelines but with informed consent from the participants, Nash has published the nude images, which admittedly did not show the participants' faces.

Upon giving a formal consent to participate in a research project, organisations often reserve the right to obtain a report of the study. It is important to note, however, that it is the researchers who assume full responsibility for the conclusions they draw and are not obliged to include the feedback they may get from the organisation in future publications. While it is now customary for organisations participating in an academic study to expect a report from the project, the time lag between data collection, analysis and the preparation of the report may at times discourage researchers, despite their earlier assurances, from sharing their results with the concerned organisations. We believe that it is important to provide research participants with the promised information and see this as part of the research project.

It is also helpful to prepare oneself for the eventuality that our research participants do not react as positively to our findings and conclusions as we would hope them to. Participants may at times find it hard to accept the conclusions drawn by the researcher. However, it is important to acknowledge that sharing our results with the research participants may also add significant value to our projects. Indeed, in some of our past projects, our participants did us a great service by taking the time to read our research reports and provide us with some additional contextual information that proved highly insightful when we started working on subsequent publications from that data set.

We witness nowadays growing pressures on researchers—also qualitative researchers—to put their data in open-access repositories to enable cross-checking and replications of the original study. Sometimes, it is a formal requirement of external funding bodies, non-negotiable by the researcher. This raises a serious issue of confidentiality—'a complex process that involves more than merely disguising the identities of research participants or sites' (Tilley and Woodthorpe 2011, p. 3). If the data are to be made available for reuse by other researchers, it is important to pay particular attention to protecting the participants of the research who should be asked to provide explicit consent to making data available after the research project is completed (see also Corti et al. 2000).

11.6   Conclusions

Ethical dilemmas faced by qualitative researchers may arise from tensions when various—often conflicting—principles meet. Although there are codes of ethics for researchers, they may only act as guidelines since research work involves dealing with unpredictable dilemmas that often require researchers to make judgement calls and to resolve them independently, on an ongoing basis, as our research evolves. We believe that many ethical dilemmas are simply insoluble; researchers often face situations where a number of principles of ethical conduct may appear to be conflicting with one another.

We agree with Wolff-Michael Roth (2005) and others (e.g. Markham and Buchanan 2012; Miller 2012) who argue that it is impossible to reduce ethics in research to an institutionalised set of top-down rules that could be applicable to all contexts. Any 'principle has to be interpreted in the light of particular situations – it is rarely if ever a matter of simply applying a rule, calculating what is best, or knowing directly what a situation requires' (Hammersley and Traianou 2012, p. 34). Research ethics needs to be regarded as an inherent element of research practice. 'Ethically important moments' (Guillemin and Gillam 2004) and ethical questions appear at every stage of the research process requiring researchers to make their own choices depending on the context of the research and, most importantly, according to their conscience. Knowledge of codes of ethics for researchers, understanding of various philosophical perspectives on which ethical postulates are based, and reflexivity, increasingly recognised as one of the key quality assurance strategies in qualitative research (Berger 2015), can all aid researchers in dealing with ethical challenges. As we have attempted to argue in this chapter, ethics is not an abstract notion or a one-off task that needs to be addressed to secure ethical approval. Foreseeing, addressing and reflecting upon ethical issues are part and parcel of everyday research practice. In light of the changing research landscape and methodological innovations, adopting 'a dialogic, case-based, inductive, and process approach to ethics' (Markham and Buchanan 2012: 5) might be more conducive to ethical decision-making that is sensitive to (at times, conflicting) contextual, cultural, institutional and legal requirements than reliance on regulatory models and procedural ethics.

References

Allen, Q. (2012). Photographs and Stories: Ethics, Benefits and Dilemmas of Using Participant Photography with Black Middle-Class Male Youth. *Qualitative Research, 12*(4), 443–458.

Beech, N., et al. (2009). "But I thought We Were Friends?" Life Cycles and Research Relationships. In S. Ybema et al. (Eds.), *Organizational Ethnography. Studying the Complexities of Everyday Life*. London: Sage.

Berger, R. (2015). Now I See it, Now I Don't: Researcher's Position and Reflexivity in Qualitative Research. *Qualitative Research, 15*(2), 219–234.

Birch, M., & Miller, T. (2002). Encouraging Participation. Ethics and Responsibilities. In M. Mauthner et al. (Eds.), *Ethics in Qualitative Research*. London: Sage.

Birch, M., Miller, T., Mauthner, M., & Jessop, J. (2012). Introduction to the Second Edition. In T. Miller, M. Birch, M. Mauthner, & J. Jessop (Eds.), *Ethics in Qualitative Research* (pp. 1–13). London: Sage.

British Sociological Association. (2006). *Visual Sociology Statement of Ethical Practice*. Retrieved from http://www.visualsociology.org.uk/about/ethical_statement.php

Cannella, G. S., & Lincoln, Y. S. (2007). Predatory vs. Dialogic Ethics. Constructing and Illusion or Ethical Practice as the Core of Research Methods. *Qualitative Inquiry, 13*(4), 315–335.

Christians, C. G. (2011). Ethics and Politics in Qualitative Research. In N. K. Denzin & Y. S. Lincoln (Eds.), *The Sage Handbook of Qualitative Research* (pp. 61–80). Los Angeles: Sage.

Clegg, J. W., & Slife, B. D. (2009). Research Ethics in the Postmodern Context. In D. M. Mertens & P. E. Ginsberg (Eds.), *The Handbook of Social Research Ethics*. London: Sage.

Corti, L., Day, A., & Backhouse, G. (2000). Confidentiality and Informed Consent. Issues for Consideration in the Preservation of Access to Qualitative Data Archives. *Forum Qualitative Sozialforschung/ Forum: Qualitative Social Research*, *1*(3).

Coupal, L. (2005). Practitioner-Research and the Regulation of Research Ethics. The Challenge of Individual, Organizational, and Social Interests. *Forum Qualitative Sozialforschung/Forum: Qualitative Social Research*, *6*(1).

Cox, S., Drew, S., Guillemin, M., Howell, C., Warr, D., & Waycott, J. (2014). *Guidelines for Ethical Visual Research Methods*. Melbourne: The University of Melbourne.

Duncombe, J., & Jessop, J. (2002). "Doing Rapport" and the Ethics of "Faking Friendship". In M. Mauthner et al. (Eds.), *Ethics in Qualitative Research*. London: Sage.

Fine, G. A. (1994). Ten Lies of Ethnography. Moral Dilemmas of Field Research. *Journal of Contemporary Ethnography*, *22*(3), 267–294.

Guillemin, M., & Gillam, L. (2004). Ethics, Reflexivity, and "Ethically Important Moments" in Research. *Qualitative Inquiry*, *10*(2), 261–280.

Haggerty, K. (2004). Ethics Creep: Governing Social Science Research in the Name of Ethics. *Qualitative Sociology*, *27*(4), 391–414.

Hammersley, M. (2009). Against the Ethicists: On the Evils of Ethical Regulation. *International Journal of Social Research Methodology*, *12*(3), 211–225.

Hammersley, M., & Traianou, A. (2012). *Ethics in Qualitative Research: Controversies and Contexts*. London: Sage.

Hedgecoe, A. (2008). Research Ethics Review and the Sociological Research Relationship. *Sociology*, *42*(5), 874–886.

Hochschild, A. (1983). *The Managed Heart: Commercialization of Human Feeling*. Berkeley: University of California Press.

Howell, C., Cox, S., Drew, S., Guillemin, M., Warr, D., & Waycott, J. (2014). Exploring Ethical Frontiers of Visual Methods. *Research Ethics*, *10*(4), 208–213.

Kara, H., & Pickering, L. (2017). New Directions in Qualitative Research Ethics. *International Journal of Social Research Methodology*, *20*(3), 239–241.

Kent, G. (2000). Ethical Principles. In D. Burton (Ed.), *Research Training for Social Scientists*. London: Sage.

Kozinets, R. (2002). The Field Behind the Screen: Using Netnography for Marketing Research in Online Communities. *Journal of Marketing Research*, *39*(1), 61–72.

Kozinets, R. (2015). *Netnography*. London: Sage.

Langer, R., & Beckman, S. C. (2005). Sensitive Research Topics: Netnography Revisited. *Qualitative Market Research: An International Journal, 8*(2), 189–203.

Lunnay, B., Borlagdan, J., McNaughton, D., & Ward, P. (2015). Ethical Use of Social Media to Facilitate Qualitative Research. *Qualitative Health Research, 25*(1), 99–109.

Markham, A., & Buchanan, E. (2012). *Ethical Decision-Making and Internet Research: Recommendations from the AoIR Ethics Working Committee (Version 2.0)*. http://aoir.org/reports/ethics2.pdf. Accessed 15 Apr 2017.

Marzano, M. (2007). Informed Consent, Deception, and Research Freedom in Qualitative Research. *Qualitative Inquiry, 13*(3), 417–436.

Mauthner, M., Birch, M., Miller, T., & Jessop, J. (2012). Conclusion: Navigating Ethical Dilemmas and New Digital Horizons. In T. Miller, M. Birch, M. Mauthner, & J. Jessop (Eds.), *Ethics in Qualitative Research*. London: Sage.

Miller, T. (2012). Reconfiguring Research Relationships: Regulation, New Technologies and Doing Ethical Research. In T. Miller, M. Birch, M. Mauthner, & J. Jessop (Eds.), *Ethics in Qualitative Research* (pp. 29–42). London: Sage.

Miller, T. (2015). Going Back: 'Stalking', Talking and Researcher Responsibilities in Qualitative Longitudinal Research. *International Journal of Social Research Methodology, 18*(3), 293–305.

Miller, T., & Bell, L. (2002). Consenting to What? Issues of Access, Gatekeeping and Informed Consent. In M. Mauthner et al. (Eds.), *Ethics in Qualitative Research*. London: Sage.

Mitchell, W., & Irvine, A. (2008). I'm Okay, You're Okay. Reflections on the Well-Being and Ethical Requirements of Researchers and Research Participants in Conducting Qualitative Fieldwork Interviews. *International Journal of Qualitative Methods, 7*(4), 31–44.

Murray, L., & Nash, M. (2016). The Challenges of Participant Photography A Critical Reflection on Methodology and Ethics in Two Cultural Contexts. *Qualitative Health Research. 22*(3): 267–294.

Neale, B. (2013). Adding Time into the Mix: Stakeholder Ethics in Qualitative Longitudinal Research. *Methodological Innovations Online, 8*(2), 6–20.

Roberts, L. D. (2015). Ethical Issues in Conducting Qualitative Research in Online Communities. *Qualitative Research in Psychology, 12*(3), 314–325.

Roth, W.-M. (2005). Ethical as Social Practice. Introducing the Debate on Qualitative Research and Ethics. *Forum Qualitative Sozialforschung/ Forum: Qualitative Social Research, 6*(1).

Sabar, G., & Sabar Ben-Yehoshua, N. (2017). I'll Sue You If You Publish My Wife's Interview': Ethical Dilemmas in Qualitative Research Based on Life Stories. *Qualitative Research*. https://doi.org/10.1177/1468794116679727.

Stanley, L., & Wise, S. (2010). The ESRC's 2010 Framework for Research Ethics: Fit for Research Purpose? *Sociological Research Online, 15*(4), 12.

Tilley, L., & Woodthorpe, K. (2011). Is it the End for Anonymity as We Know It? A Critical Examination of the Ethical Principle of Anonymity in the Context of 21st Century Demands on the Qualitative Researcher. *Qualitative Research, 11*(2), 197–212.

Van Maanen, J. (1983). The Moral Fix: On the Ethics of Fieldwork. In R. M. Emerson (Ed.), *Contemporary Field Research*. Prospect Heights: Waveland Press.

Whiteman, E. (2007). "Just Chatting" Research Ethics and Cyberspace. *International Journal of Qualitative Methods, 6*(2), 1–9.

Index

© The Author(s) 2018
M. Ciesielska, D. Jemielniak (eds.), *Qualitative Methodologies in Organization Studies*,
https://doi.org/10.1007/978-3-319-65217-7

Informed consent, 196–202, 208
Intent, 81, 114, 138, 162
Intentions, 65, 66, 69, 106, 138,
    150, 162, 201
Interpretation, x, 2, 3, 23, 29, 31,
    33, 44, 52, 61, 77–79, 83, 84,
    87, 137–154, 164
Interpretivism, 18, 119
interview
    creative, 60
    ethnographic, 76
    focused group, 20, 32, 33,
        178–180, 186
    non-standardized, 32
    non-structured, 186
    questionnaire, 2, 16
    standardized, 16
    structured, 2, 110, 186
Iteration, 42

K

Knowledge
    instrumental, 43, 204
    logico-scientific, 138, 139
    reflexive, 146, 147, 149, 150,
        152
Konecki, K., 2, 3, 33, 40, 55–57
Kostera, M., 3, 9, 13, 16
Kozinets, R., 200, 203
Kuhn, T., 8–10
Kunda, G., 3

L

Laddering
    down, 2, 50, 68
    upwards, 103, 104

Languages, 14, 24, 36, 37, 61, 76,
    80, 149, 150, 175, 178–181,
    185
    games, 80
Latour, B., 3, 115, 117, 120, 121,
    130
Learning the ropes, 100, 107
Libera. Z., 67–69
Lincoln, Y.S., 20, 21, 23, 24, 196

M

Magala, S., 49–71
Malinowski, B., 1, 149
Management
    of organizations, 1, 13, 99–111
    qualification, 102
Materialism, 11, 116
Materiality, 113
Mead, M., 149
Metaphors, 13, 14, 37, 123, 153
    epistemological, 13
Methodologies, 2, 7, 14–16, 21–23,
    25, 28, 30, 43, 53–57, 61,
    114, 116, 117, 126, 138,
    146–148, 150, 153, 154, 177,
    178, 196–198, 203
Mondragon project, 85
Morgan, G., 13–15, 17

N

Narratives, 16, 34, 35, 61–67, 93,
    111, 114–120, 128, 129, 131,
    132, 138–141
Native, 114, 115, 117, 126–131
Neo-positivism, 55
Netnography, 200

Printed by Printforce, the Netherlands